P9-ELR-619

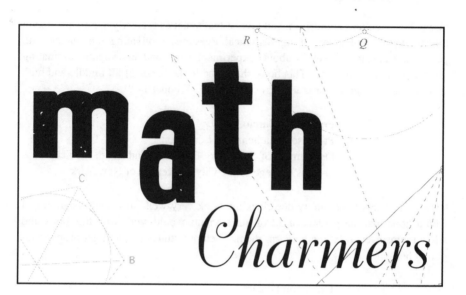

math
Charmers

"Dean Posamentier's enthusiasm and love of mathematics is clearly communicated in this marvelous collection of mathematical treasures, motivating the reader with pencil in hand, to conjecture about, experiment with, and investigate fascinating mathematical relationships. This book should be in the hands of all middle and high school students—and their mathematics teachers! Wonderful!"

Frances R. Curcio, Ph.D.
Professor, Mathematics Education
Department of Secondary Education and Youth Services
Queens College of the City University of New York

"The title of this book surely describes the book very well—it is truly charming! I would hope that the parents of all of our students would work with this book and allow themselves to let math charm them. Then our students would get proper support at home."

Dr. Anton Dobart
Director General,
Austrian Federal Ministry for Education,
Science, and Culture

"Dr. Posamentier has spent a lifetime making the subject of mathematics come to life for students and their teachers. This book is another fine tribute to the work that is possible when a brilliant mind is led by a wonderful heart. How lucky we are to add this new work to an outstanding life of achievement."

Merryl H. Tisch
Member, New York State Board of Regents

"This delightful book is just what our adult population needs: a pathway to learning about the beauty of mathematics, something that evaded most during their school days. As head of the Austrian school system, and as one who loves mathematics, I can only say that our youth would be much better off, mathematically, if more adults (i.e., parents) would acquire a love for mathematics. This book should be able to achieve that goal. I strongly recommend it to everyone!"

Elisabeth Gehrer
Austrian Federal Minister for Education, Science, and Culture

math

Charmers

Tantalizing Tidbits
for the mind

alfred s. posamentier

foreword by herbert a. hauptman, nobel laureate

Prometheus Books

59 John Glenn Drive
Amherst, New York 14228-2197

Avon Lake Public Library
32649 Electric Blvd.
Avon Lake, Ohio 44012

Published 2003 by Prometheus Books

Math Charmers: Tantalizing Tidbits for the Mind. Copyright © 2003 by Alfred S. Posamentier. All rights reserved. No part of this publication may be reproduced, stored in a retrieval system, or transmitted in any form or by any means, digital, electronic, mechanical, photocopying, recording, or otherwise, or conveyed via the Internet or a Web site without prior written permission of the publisher, except in the case of brief quotations embodied in critical articles and reviews.

Inquiries should be addressed to
Prometheus Books, 59 John Glenn Drive, Amherst, New York 14228–2197
VOICE: 716–691–0133, ext. 207; FAX: 716–564–2711
WWW.PROMETHEUSBOOKS.COM

07 06 05 04   5 4 3 2

Library of Congress Cataloging-in-Publication Data

Posamentier, Alfred S.
    Math charmers : tantalizing tidbits for the mind / Alfred S. Posamentier ; foreword by Herbert A. Hauptman.
        p. cm.
    Includes index.
    ISBN 1–59102–067–0 (pbk. : alk. paper)
    1. Mathematics—Popular works. I. Title.

QA93.P67 2003
510—dc21

2003041376

Printed in the United States of America on acid-free paper

Avon Lake Public Library
32649 Electric Blvd.
Avon Lake, Ohio 44012

In memory of my beloved parents, who, after having faced monumental adversities, provided me with the guidance to develop a love for mathematics,

and to Barbara, without whose support and encouragement this book would not have been possible.

Contents

Chapter 1 • beauty in numbers

Chapter **2** • Some **Arithmetic Marvels**

Chapter 3 • Problems with Surprising Solutions

Chapter 4 • Algebraic Entertainments

Chapter 5 • Geometric Wonders

Chapter 6 • Mathematical Paradoxes

Chapter 7 • Counting and Probability

Chapter 8 • Mathematical Potpourri

Acknowledgments

One picks up many "cute" ideas in mathematics from a variety of sources. Some ideas remain ingrained in memory while others fade with time. I have dug deeply into my memory bank to find entertaining material for this book. It is impossible to cite the many hundreds of mathematics books I have read, where I probably got some of the ideas for this book. It is also not possible to properly acknowledge the many fine colleagues and students from whom, over the past several decades, I may have gotten some of the ideas presented in this book. I would, however, like to thank Dr. Ingmar Lehmann from the Humboldt University of Berlin for some enhancement ideas he so generously offered. Thanks are also extended to Jacob Cohen, David Linker, and Amir Dagan for proofreading the manuscript of the book, and to Jan Siwanowicz for some technical assistance. Special thanks are also due to Peggy Deemer for her meticulous editing of the final manuscript. Naturally, sincere thanks are due to Barbara Lowin for her service as a sample audience, as I tried to find the most motivating ideas and then put them into appropriately intelligible form.

Alfred S. Posamentier
October 18, 2002

Foreword

bertrand Russell once wrote, "Mathematics possesses not only truth but supreme beauty, a beauty cold and austere, like that of sculpture, sublimely pure and capable of a stern perfection, such as only the greatest art can show."

Can this be the same Russell who, together with Alfred Whitehead, authored the monumental *Principia Mathematica*, which can by no means be regarded as a work of art, much less as sublimely beautiful? So what are we to believe?

Let me begin by saying that I agree completely with Russell's statement, which I first read some years ago. However, I had independently arrived at the same conviction decades earlier when, as a ten- or twelve-year-old, I first learned of the existence of the Platonic solids (these are perfectly symmetric three-dimensional figures, called polyhedra, where all faces, edges, and angles are the same—there are five such). I had been reading a book on recreational mathematics, which contained not only pictures of the five Platonic solids, but patterns, which made possible the easy construction of these polyhedra. These pictures made a profound impression on me; I could not rest until I had constructed cardboard models of all five. This was my introduction to mathematics. The Platonic solids are in fact sublimely beautiful (as Russell would say) and, at the same time, the symmetries they embody have important implications for mathematics with consequences for both geometry and algebra. In a very real sense, then, they may be regarded as providing a connecting link between geometry and algebra. Although I cannot possibly claim to have understood the full significance of this relationship some seven decades ago, I believe it fair to say that this initial encounter inspired my subsequent seventy-year love affair with mathematics.

Our next meeting is shrouded in the mists of time, but I recall with certainty that it was concerned with curves. I was so fascinated by the shape and mathematical description of a simple curve (cardioid or cissoid perhaps), which I had stumbled across in my reading, that again I could not rest until I had explored in depth as many curves as I could find in the encyclopedia during a two-month summer break. I was perhaps thirteen or fourteen at the time. I found their shapes, infinite variety, and geometric properties to be indescribably beautiful.

At the beginning of this never-to-be-forgotten summer I could not possibly have understood what was meant by the equation of a curve, which invariably appeared at the very beginning of almost every article. However, one cannot spend four or five hours a day over a two-month period without finally gaining an understanding of the relationship between a curve and its equation, between geometry and algebra, a relationship itself of profound beauty. In this way, too, I learned analytic geometry painlessly and effortlessly, in fact, with pleasure, as each curve revealed its hidden treasures—all beautiful, many profound. Is it any wonder, then, that this was a summer I shall never forget?

Why do I relate these episodes now? Because you are about to embark on a lovely book that was carefully crafted to turn you, the reader, on to mathematics. It is impossible to determine what an individual will find attractive. For me, it was symmetrically shaped solid figures and curves; for you it may be something entirely different. Yet, with the wide variety of topics and themes in this book, there will be something for everyone and hopefully much for all. Dr. Alfred S. Posamentier and I have worked on several projects together and I am well acquainted with his eagerness to demonstrate mathematics' beauty to the uninitiated. He does this with an admirable sense of enthusiasm. This is more than evident in this book, beginning with the selection of topics, which are fascinating in their own right, and taken through with his clear and comfortable presentation. He has made every effort to avoid allowing a possibly unfamiliar term or concept to slip by without defining it.

You have, therefore, in this book all the material that can evoke the beauty of mathematics presented in an accessible style—the primary goal of this book. It is the wish of every mathematician that more of society would share these beautiful morsels of mathematics with us. In my case, I took this early love for mathematics to the science research laboratories, where it provided me with insights that many scientists didn't have. This intrinsic love for mathematical structures allowed me to solve problems that stifled the chemical community for decades. I was surprisingly honored to be rewarded for my work by receiving the Nobel Prize for chemistry in 1985. I later learned that I was the first mathematician to win the Nobel Prize. All this, as a result of capturing an early love for the beauty of mathematics. Perhaps this book will open new vistas for you, where mathematics will expose its unique beauty to you. You may be pleasantly surprised in what ways this book might present new ideas or opportunities for you.

Dr. Herbert A. Hauptman
July 2002
Nobel Laureate 1985
CEO and President,
Hauptman-Woodward Medical Research Institute
Buffalo, New York

Preface

this book was inspired by the extraordinary response to a *New York Times* Op-Ed (January 2, 2002) article* I had written that called for the need to inspire people by the beauty of mathematics and not necessarily its usefulness, as is most often the case when trying to motivate youngsters to the subject. I used the year number 2,002† to motivate the reader by mentioning that it is a palindrome and then proceeded to show some entertaining aspects of a palindromic number. I could have taken it even further by having the reader take products of the number 2,002, for that, too, reveals some beautiful relationships (or quirks) of our number system. For example, look at some selected products of 2,002.

---

*See the text of the article at the end of this preface.
†Incidentally, 2002 is the product of a nice list of prime numbers: 2, 7, 11, and 13.

2,002	times	4	equals	**8,008**
2,002	times	37	equals	**74,074**
2,002	times	98	equals	**196,196**
2,002	times	123	equals	**246,246**
2,002	times	444	equals	**888,888**
2,002	times	555	equals	**1,111,110**

Following the publication of the article, I received more than five hundred letters and e-mails supporting this view and asking for ways and materials to have people see and appreciate the beauty of mathematics. I hope to be able to respond to the vast outcry for ways to demonstrate the beauty of mathematics with this book.

When I meet people socially and they discover that my field of interest is mathematics, I am usually confronted with the proud exclamation: "Oh, I was always terrible in math!" For no other subject in the curriculum would an adult be so proud about failure. Having been weak in mathematics is a badge of honor. Why is this so? Are people embarrassed to admit competence in this area? And why *are* so many people really weak in mathematics? What can be done to change this trend? Were anyone to have the definitive answer to this question, he would be the nation's education superstar. We can only conjecture where the problem lies; and then from that perspective, hope to repair it. It is my strong belief that the root of the problem lies in the inherent unpopularity of mathematics. But *why* is it so unpopular? Those who use mathematics are fine with it, but those who do not generally find it an area of study that may have caused them hardship. We must finally demonstrate the inherent beauty of mathematics, so that even those adults who do not have a daily need for it can be led to appreciate it for its beauty, and not only for its usefulness. This, then, is the objective of this book: to provide sufficient evidence of the beauty of mathematics through many examples in a variety of its branches. To make these examples attractive and effective, they were selected on the basis of the ease with which they can be understood at first reading.

Where are the societal shortcomings that lead us to this lack of popularity for mathematics? From earliest times we are told that mathematics is important to almost any endeavor we choose to pursue.

When a young child is encouraged to do well in school in mathematics, it is usually accompanied with "You'll need mathematics if you want to be a _____." For the young child this is a useless justification since his career goals are not yet of any concern to him. Thus, this is an empty statement. Sometimes a child is told to do better in mathematics or else_____." This, too, does not have a lasting effect on the child, who does just enough to avoid punishment. He will give mathematics attention only to avoid further difficulty from his parents.

To compound this lack of popularity of mathematics among the populace, the child who may not be doing as well in mathematics as in other subject areas is consoled by his parents by being told that they, too, were not good in mathematics in their school days. This negative role model can have a most deleterious effect on a youngster's motivation toward mathematics.

For school administrators, performance in mathematics will typically be the bellwether for their schools' success or weakness. When their schools perform well either in comparison to a national standard or in comparison to neighboring school districts, then they breathe a sigh of relief. On the other hand, when their schools do not perform well, there is immediate pressure to fix the situation. More often than not, these schools place the blame on the teachers. Usually, a "quick fix" in-service training program is initiated for the math teachers in the schools. Unless the in-service training program is carefully tailored to the particular teachers, little can be expected in the way of improved student performance. Very often, a school or district will blame the curriculum (or textbook) and then alter it in the hope of bringing about immediate change. This can be dangerous, since a sudden change in curriculum can leave the teachers ill prepared for this new material and thereby cause further difficulty. When an in-service training program purports to have the "magic formula" to improve teacher performance, one ought to be a bit suspicious. Making teachers more effective requires a considerable amount of effort spread over a long time.* Then it is an extraordinarily difficult task for a number of rea-

*Toward this end, a companion version of this book is being published under the title *Math Wonders to Inspire Teachers and Students* by the Association for Supervision and Curriculum Development.

sons. First, one must clearly determine where the weaknesses lie. Is it a general weakness in content? Are the pedagogical skills lacking? Are the teachers simply lacking motivation? Or is it a combination of these factors? Whatever the problem, it is generally not shared by every math teacher in the school. This then implies that a variety of in-service training programs would need to be instituted to meet the overall weakness of instruction. This is rarely, if ever, done because of the organizational and financial considerations of providing in-service training on an individual basis. The problem of making mathematics instruction more successful by changing the teachers' performance is clearly only a part of the solution.

International comparative studies constantly place our country's schools at a relatively low ranking. Thus, politicians take up the cause of raising mathematics performance. They wear the hat of "education president," "education governor," or "education mayor" and authorize new funds to meet educational weaknesses. These funds are usually spent to initiate professional development in the form of the in-service training programs we just discussed. Their effectiveness is questionable at best for the reasons outlined above.

What, then, remains for us to do to improve the mathematics performance of youngsters in the schools? The society as a whole must embrace mathematics as an area of beauty (and fun) and not merely as a useful subject, without which further study in many areas would not be possible (although this latter statement may be true). We must begin with the parents, who as adults already have their minds made up on their feelings about mathematics. Although it is a difficult task to turn on an adult to mathematics when he already is negatively disposed to the subject, the goal of this book is to do precisely that. The question that still remains is how best to achieve this goal.

Someone not particularly interested in mathematics, or someone fearful of the subject, must be presented with illustrations that are extremely easy to comprehend. He needs to be presented with examples that do not require much explanation, ones that sort of "bounce off the page" in their attractiveness. It is also helpful if the examples are largely visual. They can be—but don't have to be—recreational in nature. Above all, they should elicit the "wow" response, that feeling that there really is something special about the nature of mathematics. This specialness can manifest itself in a number of ways. It can be a simple problem, where mathematical reasoning leads to an unexpectedly simple (or elegant) solution. It may be an illustration of the nature of numbers that leads to a "gee whiz" reaction. It may be a geometrical relationship that intuitively seems implausible. Probability also has some such entertaining phenomena that can evoke such responses. Whatever the illustration, the result must be quickly and efficiently obtained. With enough of these illustrations, the reader should begin to form a more positive feeling about mathematics.

At the point that such a turnaround of feelings occurs, the adult usually asks, "Why wasn't I shown these lovely things when I was in school?" We can't answer that and we can't change that. We can, however, make more adults goodwill ambassadors for mathematics and make them more resourceful so that they introduce these mathematics charmers to others—most notably, young people. So we must all, perhaps using these mathematics charmers, change the societal perception of mathematics to one where it is admired and appreciated. Only then will we bring about meaningful change in mathematics achievement, in addition to an appreciation for the beauty of mathematics.

The following article first appeared in the *New York Times* on January 2, 2002.

Madam, I'm 2002 — a Numerically Beautiful Year

David Goldin*

Now, in the year 2002, we are the last generation for more than the next thousand years to experience two palindromic years in a lifetime. Palindromes in mathematics are numbers that read the same in both directions, like 2002 and 1991. In the English language there are very well known palindromic words, like "rotator" and "reviver," and there are also sentences that are palindromic, like "Madam, I'm Adam."

Palindromic numbers are not only pleasing in appearance, but also harbor some nice, curious qualities. For example, take any number, write it in reverse order, and add the two numbers. The sum will likely be a palindromic number, and if not, then simply continuing this process by adding the sum to its reverse should eventually lead to a palindromic number. This surprising result is the kind of discovery that can pique interest in numbers even in students indifferent to math.

Teachers should be able to capitalize on the beauty in mathematics, and specifically the charm of some numbers to hook students on studying mathematics. But qualified math teachers who might actually inspire children are in short supply, and math teaching in today's schools is often dry and boring. The problem is not new. Math instruction at the elementary school level, when students

---

*The artist David Goldin can be contacted at davidgoldin@interport.net.

form their first impressions about the nature of math and their own abilities with numbers, has historically been mediocre. Without a strong beginning, a student's chances for sustaining interest in this field are very small indeed. The problem is compounded by relatively weak teaching at the secondary level as well.

A number of quick-fix programs to increase the supply of secondary school math teachers have been put in place around the country through the process of alternative certification. Participants in one such program in New York City function reasonably well, but many also lack the ability to bring to their classes a depth of understanding of math that allows a teacher to do more than merely conform to the prescribed instructional plan. A marginally competent teacher may not necessarily be skilled enough to promote interest in or appreciation for math among students.

At a time when there is a national shortage of math teachers, made worse by a low supply of math-prepared students, we must look beyond the quick-fix solutions. We must develop better and more creative training programs for elementary math teachers. We need to give them more classroom time on this subject. And when students begin to pursue the study of math, we must make the teaching profession more attractive, financially as well as by giving teachers more control of how they teach.

The point is to make math intrinsically interesting to children. We should not have to sell mathematics by pointing to its usefulness in other subject areas, which, of course, is real. Love for math will not come about by trying to convince a child that it happens to be a handy tool for life. It grows when a good teacher can draw out a child's curiosity about how numbers and mathematical principles work. The very high percentage of adults who are unashamed to say that they are bad with math is a good indication of how maligned the subject is and how very little we were taught in school about the enchantment of numbers. ∎

1

Beauty in Numbers

We are accustomed to seeing numbers in charts and tables on the sports or business pages of a newspaper. We use numbers continuously in our everyday life experiences, either to represent a quantity or to designate something such as a street, address, or page. We use numbers without ever taking the time to observe some of their unusual properties. That is, we don't stop to smell the flowers as we walk through a garden, or as it is more commonly said: "take time to smell the roses." Inspecting some of these unusual number properties provides us with a much deeper appreciation for these symbols that we all too often take for granted.

There are basically two types of number properties, those that are "quirks" of the decimal system and those that are true in any number system. Naturally, the latter gives us better insight into mathematics, while the former merely points out the arbitrary nature of using a decimal system. One might ask why we use a decimal system when today we find the foundation of computers relies on a binary system.* The answer is clearly historical, and no doubt emanates from our number of fingers.

On the surface the two types of peculiarities do not differ much in their appearance, just their justification. Since this book is not intended for the mathematician (although they, too, might find some little gems here), the justifications or explanations will be kept simple and intelligible to the general reader. By the same token, in some cases the explanation might lead the reader to further research into or inspection of the phenomenon. The moment you are brought to the point where you question *why* the property exhibited occurred, you're hooked! That is the goal of this chapter, to make you want to marvel at the results and question them. Although the explanations may leave you with some questions, you will be well on your way to doing some individual explorations. That is when you really get to appreciate the mathematics involved. It is during these "private" investigations that genuine learning takes place. Don't thwart it; encourage it!

Above all, take note of the beauty of the number relationships. Without further talk, let's go to the delights in the realm of numbers and number relationships.

I.I. Surprising Number Patterns I

Perhaps one of the easiest ways to begin to appreciate the beauty of mathematics is to observe some of the enchanting relationships that it holds. The following arithmetic symmetries speak for themselves. There are times when the charm of mathematics lies in the surprising nature of its number system. Look and enjoy!†

---

*Where the decimal system gives the value to a number by assigning powers of 10 (right to left) to the digits: 1, 10, 100, 1,000, which, as powers of 10, are written as 10^0, 10^1, 10^2, and 10^3, the binary system is one where the place values of the numbers (from right to left) are successive powers of 2, and the symbols used are 0 and 1. The binary number 11010 has a value $1(2^4) + 1(2^3) + 0(2^2) + 1(2^1) + 0(2^0) = 26$.

†We are using the dot (•) to represent multiplication, since that is the symbol most universally used to indicate multiplication.

$$1 \bullet 1 = 1$$
$$11 \bullet 11 = 121$$
$$111 \bullet 111 = 12{,}321$$
$$1{,}111 \bullet 1{,}111 = 1{,}234{,}321$$
$$11{,}111 \bullet 11{,}111 = 123{,}454{,}321$$
$$111{,}111 \bullet 111{,}111 = 12{,}345{,}654{,}321$$
$$1{,}111{,}111 \bullet 1{,}111{,}111 = 1{,}234{,}567{,}654{,}321$$
$$11{,}111{,}111 \bullet 11{,}111{,}111 = 123{,}456{,}787{,}654{,}321$$
$$111{,}111{,}111 \bullet 111{,}111{,}111 = 12{,}345{,}678{,}987{,}654{,}321$$

Notice that all these products are palindromes. Now for another cute pattern.

$$1 \bullet 8 + 1 = 9$$
$$12 \bullet 8 + 2 = 98$$
$$123 \bullet 8 + 3 = 987$$
$$1{,}234 \bullet 8 + 4 = 9{,}876$$
$$12{,}345 \bullet 8 + 5 = 98{,}765$$
$$123{,}456 \bullet 8 + 6 = 987{,}654$$
$$1{,}234{,}567 \bullet 8 + 7 = 9{,}876{,}543$$
$$12{,}345{,}678 \bullet 8 + 8 = 98{,}765{,}432$$
$$123{,}456{,}789 \bullet 8 + 9 = 987{,}654{,}321$$

Sometimes the beauty is a bit camouflaged. Notice (below) how various products of 76,923 yield numbers in the same order but with a different starting point. In each succeeding row the first digit of the product goes to the end of the number to form the next product. Otherwise, the order of the digits remains intact.

$$76{,}923 \bullet 1 = 076{,}923$$
$$76{,}923 \bullet 10 = 769{,}230$$
$$76{,}923 \bullet 9 = 692{,}307$$
$$76{,}923 \bullet 12 = 923{,}076$$
$$76{,}923 \bullet 3 = 230{,}769$$
$$76{,}923 \bullet 4 = 307{,}692$$

In the multiplications that follow, the various products of 76,923 yield different numbers from those above. The results are in the same order but with a different starting point. Again, the first digit of the product goes to the end of the number to form the next product. Otherwise, the order of the digits remains intact.

76,923	•	2	=	153,846
76,923	•	7	=	538,461
76,923	•	5	=	384,615
76,923	•	11	=	846,153
76,923	•	6	=	461,538
76,923	•	8	=	615,384

Another peculiar number is 142,857. When it is multiplied by the numbers 2 through 8, the results are astonishing. Consider the following products and describe the peculiarity.

142,857	•	2	=	285,714
142,857	•	3	=	428,571
142,857	•	4	=	571,428
142,857	•	5	=	714,285
142,857	•	6	=	857,142

You can see symmetries in the products, but notice also that the same digits are used in the product as in the first factor. Furthermore, consider the order of the digits. With the exception of the starting point, they are in the same sequence. These are just some lovely peculiarities of the number system we use. When we continue with these products of 142,857, we get some other weird results. Look at the product, $142,857 • 7 = 999,999$. Where did this come from?

It gets even stranger with this product: $142,857 • 8 = 1,142,856$. If we remove the millions digit and add it to the units digit, the original number is formed (i.e., $142,856 + 1 = 142,857$).

These are just a few numbers that yield strange products. Yet in an intriguing way, they evoke some of the beauty in mathematics.

1.2. Surprising Number Patterns II

Here are some more peculiarities of mathematics that depend on the nature of our number system, more evidence that mathematics has some hidden wonders. Again, not many words are needed to demonstrate the charm, for it is obvious at first sight. Just look, enjoy, and share these amazing properties with your friends. Others may need encouragement! Let them appreciate the patterns and if possible try to look for an "explanation" for this.

$$12,345,679 \cdot 9 = 111,111,111$$
$$12,345,679 \cdot 18 = 222,222,222$$
$$12,345,679 \cdot 27 = 333,333,333$$
$$12,345,679 \cdot 36 = 444,444,444$$
$$12,345,679 \cdot 45 = 555,555,555$$
$$12,345,679 \cdot 54 = 666,666,666$$
$$12,345,679 \cdot 63 = 777,777,777$$
$$12,345,679 \cdot 72 = 888,888,888$$
$$12,345,679 \cdot 81 = 999,999,999$$

In the following pattern chart, notice that the first and last digits of the products are the digits of the multiples of 9.

$$987,654,321 \cdot 9 = 08\ 888\ 888\ 889$$
$$987,654,321 \cdot 18 = 17\ 777\ 777\ 778$$
$$987,654,321 \cdot 27 = 26\ 666\ 666\ 667$$
$$987,654,321 \cdot 36 = 35\ 555\ 555\ 556$$
$$987,654,321 \cdot 45 = 44\ 444\ 444\ 445$$
$$987,654,321 \cdot 54 = 53\ 333\ 333\ 334$$
$$987,654,321 \cdot 63 = 62\ 222\ 222\ 223$$
$$987,654,321 \cdot 72 = 71\ 111\ 111\ 112$$
$$987,654,321 \cdot 81 = 80\ 000\ 000\ 001$$

The number patterns give you further evidence of the beauty in mathematics.

I.3. Surprising Number Patterns III

Again, not many words are needed to demonstrate the charm, for it is obvious at first sight. Some of these leave you a bit baffled. But remember: We are only beginning to get you into the mood to appreciate a view of mathematics you may not have been exposed to before. Inspect these and just enjoy them.

$$0 \cdot 9 + 1 = 1$$
$$1 \cdot 9 + 2 = 11$$
$$12 \cdot 9 + 3 = 111$$
$$123 \cdot 9 + 4 = 1,111$$
$$1,234 \cdot 9 + 5 = 11,111$$
$$12,345 \cdot 9 + 6 = 111,111$$
$$123,456 \cdot 9 + 7 = 1,111,111$$
$$1,234,567 \cdot 9 + 8 = 11,111,111$$
$$12,345,678 \cdot 9 + 9 = 111,111,111$$

A similar process yields another interesting pattern.

$$0 \cdot 9 + 8 = 8$$
$$9 \cdot 9 + 7 = 88$$
$$98 \cdot 9 + 6 = 888$$
$$987 \cdot 9 + 5 = 8,888$$
$$9,876 \cdot 9 + 4 = 88,888$$
$$98,765 \cdot 9 + 3 = 888,888$$
$$987,654 \cdot 9 + 2 = 8,888,888$$
$$9,876,543 \cdot 9 + 1 = 88,888,888$$
$$98,765,432 \cdot 9 + 0 = 888,888,888$$

Now a logical thing to inspect would be the pattern of these strange products.

1 •	8	=	8
11 •	88	=	968
111 •	888	=	98,568
1,111 •	8,888	=	9,874,568
11,111 •	88,888	=	987,634,568
111,111 •	888,888	=	98,765,234,568
1,111,111 •	8,888,888	=	9,876,541,234,568
11,111,111 •	88,888,888	=	987,654,301,234,568
111,111,111 •	888,888,888	=	98,765,431,901,234,568
1,111,111,111 •	8,888,888,888	=	9,876,543,207,901,234,568

How might you describe this pattern?

1.4. Surprising Number Patterns IV

Again, words would spoil the effect that the following number patterns project. Yet in this case, you will notice that much is dependent on the number 1,001, which is the product of 7, 11, and 13. Furthermore, when we multiply 1,001 by a three-digit number, the result is nicely symmetric. For example, 987 • 1,001 = 987,987. Thus, reversing this relationship, any six-digit number comprised of two repeating sequences of three digits is divisible by 7, 11, and 13. For example,

$$\frac{643,643}{7} = 91,949$$
$$\frac{643,643}{11} = 58,513$$
$$\frac{643,643}{13} = 49,511$$

We can also draw another conclusion from this interesting number 1,001, that is, a number with six repeating digits is always divisible by 3, 7, 11, and 13.

$$\frac{111,111}{3} = 37,037$$
$$\frac{111,111}{7} = 15,873$$
$$\frac{111,111}{11} = 10,101$$
$$\frac{111,111}{13} = 8,547$$

You may wish to explore this symmetric number a bit more to search for other relationships that might be found.

I.5. Surprising Number Patterns V

Here are some more examples of the beauty of mathematics that depends on the peculiar nature of its number system. Again, the patterns speak for themselves. These depend on the property described in the last section and the unusual property of the number 9.

$$
\begin{aligned}
999{,}999 \cdot 1 &= 0{,}999{,}999 \\
999{,}999 \cdot 2 &= 1{,}999{,}998 \\
999{,}999 \cdot 3 &= 2{,}999{,}997 \\
999{,}999 \cdot 4 &= 3{,}999{,}996 \\
999{,}999 \cdot 5 &= 4{,}999{,}995 \\
999{,}999 \cdot 6 &= 5{,}999{,}994 \\
999{,}999 \cdot 7 &= 6{,}999{,}993 \\
999{,}999 \cdot 8 &= 7{,}999{,}992 \\
999{,}999 \cdot 9 &= 8{,}999{,}991 \\
999{,}999 \cdot 10 &= 9{,}999{,}990
\end{aligned}
$$

Again the number 9 presents some nice peculiarities.*

$$
\begin{aligned}
9 \cdot 9 &= 81 \\
99 \cdot 99 &= 9{,}801 \\
999 \cdot 999 &= 998{,}001 \\
9{,}999 \cdot 9{,}999 &= 99{,}980{,}001 \\
99{,}999 \cdot 99{,}999 &= 9{,}999{,}800{,}001 \\
999{,}999 \cdot 999{,}999 &= 999{,}998{,}000{,}001 \\
9{,}999{,}999 \cdot 9{,}999{,}999 &= 99{,}999{,}980{,}000{,}001
\end{aligned}
$$

While playing with the number 9, you might try to find an eight-digit number in which no digit is repeated and which, when multiplied by 9, yields a nine-digit number in which no digit is repeated.

---

*Some properties of the number 9 are due to the fact that it is 1 less than the base, 10.

Here are a few such possibilities:

$$81,274,365 \bullet 9 = 731,469,285$$
$$72,645,831 \bullet 9 = 653,812,479$$
$$58,132,764 \bullet 9 = 523,194,876$$
$$76,125,483 \bullet 9 = 685,129,347$$

I.6. Surprising Number Patterns VI

This delightful arrangement of numbers plays on a certain symmetry that is merely embellished at the right. The nice part is to be found in the triangular array. Perhaps you can develop your own mathematically beautiful arrangement of number patterns.

$$1 = 1 \qquad\qquad = 1 \bullet 1 = 1^2$$
$$1+2+1 = 2+2 \qquad\qquad = 2 \bullet 2 = 2^2$$
$$1+2+3+2+1 = 3+3+3 \qquad\qquad = 3 \bullet 3 = 3^2$$
$$1+2+3+4+3+2+1 = 4+4+4+4 \qquad\qquad = 4 \bullet 4 = 4^2$$
$$1+2+3+4+5+4+3+2+1 = 5+5+5+5+5 \qquad\qquad = 5 \bullet 5 = 5^2$$
$$1+2+3+4+5+6+5+4+3+2+1 = 6+6+6+6+6+6 \qquad\qquad = 6 \bullet 6 = 6^2$$
$$1+2+3+4+5+6+7+6+5+4+3+2+1 = 7+7+7+7+7+7+7 \qquad\qquad = 7 \bullet 7 = 7^2$$
$$1+2+3+4+5+6+7+8+7+6+5+4+3+2+1 = 8+8+8+8+8+8+8+8 \qquad\qquad = 8 \bullet 8 = 8^2$$
$$1+2+3+4+5+6+7+8+9+8+7+6+5+4+3+2+1 = 9+9+9+9+9+9+9+9+9 = 9 \bullet 9 = 9^2$$

I.7. Amazing Power Relationships

Our number system has many unusual features built into it. Discovering them can certainly be a rewarding experience. Sometimes we stumble onto these relationships and other times they are the result of experimentation and avid searching—based on a hunch. The famous mathematician Carl Friedrich Gauss (1777–1855) discovered a fair number of relationships (which he later proved to establish theorems)[*] on the basis of his superior arithmetic abilities.

---

[*]Theorems are statements that have been proved true.

Consider the following relationship and describe what is going on.

$$81 = (8 + 1)^2 = 9^2$$

We took the square of the sum of the digits. Now look below:

$$4{,}913 = (4 + 9 + 1 + 3)^3 = 17^3$$

In both cases, we have taken the sum of the digits of this number to a power and ended up with the number we started with. Impressed? You ought to be, for this is quite astonishing. Now, to find other such numbers is no mean feat.

The list below will provide you with lots of examples of these unusual numbers. Enjoy yourself!

Number	(Sum of the Digits)^n	Number	(Sum of the Digits)^n
81 =	9^2	34,012,224 =	18^6
		8,303,765,625 =	45^6
512 =	8^3	24,794,911,296 =	54^6
4,913 =	17^3	68,719,476,736 =	64^6
5,832 =	18^3		
17,576 =	26^3	612,220,032 =	18^7
19,683 =	27^3	10,460,353,203 =	27^7
		27,512,614,111 =	31^7
2,401 =	7^4	52,523,350,144 =	34^7
234,256 =	22^4	271,818,611,107 =	43^7
390,625 =	25^4	1,174,711,139,837 =	53^7
614,656 =	28^4	2,207,984,167,552 =	58^7
1,679,616 =	36^4	6,722,988,818,432 =	68^7
17,210,368 =	28^5	20,047,612,231,936 =	46^8
52,521,875 =	35^5	72,301,961,339,136 =	54^8
60,466,176 =	36^5	248,155,780,267,521 =	63^8
205,962,976 =	46^5		
20,864,448,472,975,628,947,226,005,981,267,194,447,042,584,001 =			207^{20}

The beauty is self-evident!

1.8. beautiful number relationships

Who said numbers can't form beautiful relationships? Some of these unique situations might give you the feeling that there is more to "numbers" than meets the eye. You are encouraged not only to verify these relationships, but also to find others that can be considered "beautiful."

Let's look at the numbers 135 and 175. On the surface you would think they have nothing in common. Well, they are two of the numbers that share a very special property. Look what happens when each of their digits is raised to a power one greater than that of the previous digit.

Notice that each of these numbers equals the sum of its digits raised to consecutive exponents.

$$135 = 1^1 + 3^2 + 5^3$$
$$175 = 1^1 + 7^2 + 5^3$$
$$518 = 5^1 + 1^2 + 8^3$$
$$598 = 5^1 + 9^2 + 8^3$$

It is natural to ask if there are four-digit numbers that also have this amazing property. Here are some that satisfy this relationship.

$$1,306 = 1^1 + 3^2 + 0^3 + 6^4$$
$$1,676 = 1^1 + 6^2 + 7^3 + 6^4$$
$$2,427 = 2^1 + 4^2 + 2^3 + 7^4$$

Now if you thought these were unusual numbers, you will probably be quite enchanted with the next unusual number properties. They are really amazing. Notice the relationship between the exponents and the numbers.*

$$3,435 = 3^3 + 4^4 + 3^3 + 5^5$$
$$438,579,088 = 4^4 + 3^3 + 8^8 + 5^5 + 7^7 + 9^9 + 0^0 + 8^8 + 8^8$$

Who said mathematics doesn't have its "beauties" to show off?

---

*In the second example, the expression 0^0 is defined by mathematicians to be indeterminate, yet for simplicity's sake (and to make our example work) we shall give it a value of 0.

I.9. Unusual Number Relationships

Every so often we stumble on some relatively useless information that still contains an element of beauty resulting from its peculiarity. One such is the unusual relationship between numbers that we will exhibit here. There is not much explanation for them. Just enjoy them and see if you can find others.

What we have are pairs of numbers where the product and the sum are reversals of each other.

The two numbers		their product	their sum
9	9	81	18
3	24	72	27
2	47	94	49
2	497	994	499

Can you find another pair of numbers that exhibits this unusual property?

Now on a separate note, we have another strange relationship that, within its unusualness, gives us a sense that there can be some real beauty in mathematics. We just have to find it.

Here we have a symmetry with numbers and the sum of the factorials* of their digits.

$$1 = 1!$$
$$2 = 2!$$
$$145 = 1! + 4! + 5!$$
$$40,585 = 4! + 0! + 5! + 8! + 5! \text{ [Remember that } 0! = 1]$$

That appears to be all of this sort that exists, so don't bother looking for others.

---

*The factorial notation "!" represents the product of all the integers equal to and less than the number with the factorial notation. For example, $5! = (5)(4)(3)(2)(1) = 120$, $8! = (8)(7)(6)(5)(4)(3)(2)(1) = 40,320$, and $n! = (n)(n-1)(n-2)(n-3) \dots (3)(2)(1)$. Notice here we are using the parentheses to indicate multiplication.

1.10. Strange Equalities

When we speak of unusual or strange relationships, those that follow must rank high on the list. Yet in this case, strange can be enchanting. The reader will probably want to know why this happens. What is it about these numbers that allows this relationship to occur? There are no clever answers to this question. All we can do is find and limit the possibilities. There are times when the numbers speak more effectively than any explanation. Here is one such case. Just look at these equalities and enjoy them for what they are: Unusual!

$$1^1 + 6^1 + 8^1 = 15 = 2^1 + 4^1 + 9^1$$
$$1^2 + 6^2 + 8^2 = 101 = 2^2 + 4^2 + 9^2$$

$$1^1 + 5^1 + 8^1 + 12^1 = 26 = 2^1 + 3^1 + 10^1 + 11^1$$
$$1^2 + 5^2 + 8^2 + 12^2 = 234 = 2^2 + 3^2 + 10^2 + 11^2$$
$$1^3 + 5^3 + 8^3 + 12^3 = 2{,}366 = 2^3 + 3^3 + 10^3 + 11^3$$

$$1^1 + 5^1 + 8^1 + 12^1 + 18^1 + 19^1 = 63 = 2^1 + 3^1 + 9^1 + 13^1 + 16^1 + 20^1$$
$$1^2 + 5^2 + 8^2 + 12^2 + 18^2 + 19^2 = 919 = 2^2 + 3^2 + 9^2 + 13^2 + 16^2 + 20^2$$
$$1^3 + 5^3 + 8^3 + 12^3 + 18^3 + 19^3 = 15{,}057 = 2^3 + 3^3 + 9^3 + 13^3 + 16^3 + 20^3$$
$$1^4 + 5^4 + 8^4 + 12^4 + 18^4 + 19^4 = 260{,}755 = 2^4 + 3^4 + 9^4 + 13^4 + 16^4 + 20^4$$

Not much can be said here. Perhaps "wow" would be in order. The beauty is in the unusual!

1.11. the amazing number 1,089

This is about a number that has some truly exceptional properties. We begin by showing how the number 1,089 just happens to "pop up" when least expected and then we'll take another look at this number.

We shall begin by having *you* select any three-digit number, where the units and hundreds digit are not the same, and follow the instructions below.

Follow these instructions step by step, while we do it below each instruction.

**Choose any three-digit number
(where the units and hundreds digit are not the same).**

We will do it with you here by arbitrarily selecting: **825**

Reverse the digits of this number you have selected.

We will continue here by reversing the digits of 825 to get: **528**

**Subtract the two numbers
(naturally, the larger minus the smaller).**

Our calculated difference is: $825 - 528 = \mathbf{297}$

Once again, reverse the digits of this difference.

Reversing the digits of 297 we get the number: **792**

Now, add your last two numbers.

We then add the last two numbers to get: $297 + 792 = \mathbf{1,089}$

**Your result should be the same as ours even though
your starting number was different from ours. ***

You will probably be astonished that regardless of which number you selected at the beginning, you got the same result as we did, 1,089.

How does this happen? Is this a "freak property" of this number? Did we do something devious in our calculations?

Unlike the previous unit, which depended on a peculiarity of the decimal system, this illustration of a mathematical oddity depends on the operations. Before we explore (for the more motivated reader) why this happens, let yourself be impressed with a further property of this lovely number 1,089.

---

*If not, then you made a calculation error. Check it.

Let's look at the first nine multiples of 1,089.

$$1,089 \bullet 1 = 1,089$$
$$1,089 \bullet 2 = 2,178$$
$$1,089 \bullet 3 = 3,267$$
$$1,089 \bullet 4 = 4,356$$
$$1,089 \bullet 5 = 5,445$$
$$1,089 \bullet 6 = 6,534$$
$$1,089 \bullet 7 = 7,623$$
$$1,089 \bullet 8 = 8,712$$
$$1,089 \bullet 9 = 9,801$$

Do you notice a pattern among the products? Look at the first and ninth products (i.e., 1,089 and 9,801). They are the reverses of one another. The second and the eighth products (i.e., 2,178 and 8,712) are also reverses of one another. And so the pattern continues, until the fifth product, 5,445, is the reverse of itself, known as a palindromic number.*

Notice in particular that $1,089 \bullet 9 = 9,801$, which is the reversal of the original number. The same property holds for $10,989 \bullet 9 = 98,901$, and similarly, $109,989 \bullet 9 = 989,901$.

You should recognize that we altered the original number, 1,089, by inserting a 9 in the middle of the number to get 10,**9**89, and extended that by inserting 99 in the middle of the number 1,089 to get 10**9**,**9**89. It would be nice to conclude from this that each of the following numbers has the same property: 1,0**99**,**9**89, 10,**999**,**9**89, 10**9**,**999**,**9**89, 1,0**99**,**999**,**9**89, 10,**999**,**999**,**9**89, and so on.

As a matter of fact, there is only one other number with four or fewer digits, where a multiple of itself is equal to its reversal, and that is the number 2,178 (which just happens to be 2 • 1,089), since 2,178 • 4 = 8,712. Wouldn't it be nice if we could extend this as we did with the above example by inserting 9s into the middle of the number to generate other numbers that have the same property? Yes, it is true that

*We have more about palindromic numbers in section 1.16.

21,978 • 4 = 87,912
219,978 • 4 = 879,912
2,199,978 • 4 = 8,799,912
21,999,978 • 4 = 87,999,912
219,999,978 • 4 = 879,999,912
2,199,999,978 • 4 = 8,799,999,912
and so on.

As if the number 1,089 didn't already have enough cute properties, here is another one that (sort of) extends the 1,089: We will actually consider the number 1,089 in two parts: the numbers 1 and 89.

Let's see what happens when you take any number and get the sum of the squares of its digits. Then continue this process of finding the sum of the squares of the digits. Each time, curiously enough, you will eventually reach 1 or 89. Take a look at some examples that follow.

We will begin with the number 30. So we can say that $n = 30$, and we will find the sum of the squares of the digits of this number:

$$3^2 + 0^2 = 9, 9^2 = 81, 8^2 + 1^2 = 65, 6^2 + 5^2 = 61, 6^2 + 1^2 = 37, 3^2 + 7^2 = 58, 5^2 + 8^2 = \mathbf{89}, 8^2 + 9^2 = 145, 1^2 + 4^2 + 5^2 = 42, 4^2 + 2^2 = 20, 2^2 + 0^2 = 4, 4^2 = 16, 1^2 + 6^2 = 37, 3^2 + 7^2 = 58, 5^2 + 8^2 = \mathbf{89}, ...$$

Once we reached 89, we got into what we call a loop, since we always seem to get back to the number 89, when we repeat the process. Let's try this with the number 31.

So we will let $n = 31$: $3^2 + 1^2 = 10, 1^2 + 0^2 = \mathbf{1}, 1^2 = \mathbf{1}$

Again, for the number 1 a loop is formed, getting us back to 1 over and over.

We shall now try 32, and we let $n = 32$: $3^2 + 2^2 = 13, 1^2 + 3^2 = 10, 1^2 + 0^2 = \mathbf{1}, 1^2 = \mathbf{1}$

For $n = 33$: $3^2 + 3^2 = 18, 1^2 + 8^2 = 65, 6^2 + 5^2 = 61, 6^2 + 1^2 = 37, 3^2 + 7^2 = 58, 5^2 + 8^2 = \mathbf{89}, 8^2 + 9^2 = 145, 1^2 + 4^2 + 5^2 = 42, 4^2 + 2^2 = 20, 2^2 + 0^2 = 4, 4^2 = 16, 1^2 + 6^2 = 37, 3^2 + 7^2 = 58, 5^2 + 8^2 = \mathbf{89}, ...$

For $n = 80$: $8^2 + 0^2 = 64$, $6^2 + 4^2 = 52$, $5^2 + 2^2 = 29$, $2^2 + 9^2 = 85$, $8^2 + 5^2 = \mathbf{89}$, $8^2 + 9^2 = 145$, $1^2 + 4^2 + 5^2 = 42$, $4^2 + 2^2 = 20$, $2^2 + 0^2 = 4$, $4^2 = 16$, $1^2 + 6^2 = 37$, $3^2 + 7^2 = 58$, $5^2 + 8^2 = \mathbf{89}$, ...

For $n = 81$: $8^2 + 1^2 = 65$, $6^2 + 5^2 = 61$, $6^2 + 1^2 = 37$, $3^2 + 7^2 = 58$, $5^2 + 8^2 = \mathbf{89}$, $8^2 + 9^2 = 145$, $1^2 + 4^2 + 5^2 = 42$, $4^2 + 2^2 = 20$, $2^2 + 0^2 = 4$, $4^2 = 16$, $1^2 + 6^2 = 37$, $3^2 + 7^2 = 58$, $5^2 + 8^2 = \mathbf{89}$, ...

For $n = 82$: $8^2 + 2^2 = 68$, $6^2 + 8^2 = 100$, $1^2 + 0^2 + 0^2 = \mathbf{1}$, $1^2 = \mathbf{1}$

For $n = 85$: $8^2 + 5^2 = \mathbf{89}$, $8^2 + 9^2 = 145$, $1^2 + 4^2 + 5^2 = 42$, $4^2 + 2^2 = 20$, $2^2 + 0^2 = 4$, $4^2 = 16$, $1^2 + 6^2 = 37$, $3^2 + 7^2 = 58$, $5^2 + 8^2 = \mathbf{89}$, ...

Now let's go back to the original oddity of the number 1,089, the one where we used digit reversals in order to generate 1,089 from a selected three-digit number. We assumed that any number we chose would lead us to 1,089. How can we be sure? Well, we could try all possible three-digit numbers to see if it works. That would be tedious and not particularly elegant. An investigation of this oddity requires nothing more than some knowledge of elementary algebra. For the reader who might be curious about this phenomenon, we will provide an algebraic explanation as to why it "works."

We shall represent the arbitrarily selected three-digit number, *htu* as $100h + 10t + u$, where h represents the hundreds digit, t represents the tens digit, and u represents the units digit.

Let $h > u$,* which would be the case either in the number you selected or the reverse of it.

In the subtraction, $u - h < 0$; therefore, take 1 from the tens place (of the minuend) making the units place $10 + u$.

Since the tens digits of the two numbers to be subtracted are equal, and 1 was taken from the tens digit of the minuend, then the value of this digit is $10(t - 1)$. The hundreds digit of the minuend is $h - 1$, because 1 was taken away to enable subtraction in the tens place, making the value of the tens digit $10(t - 1) + 100 = 10(t + 9)$.

We can now do the first subtraction:

---

*The symbol > means "greater than" and the symbol < means "less than."

$$100(h-1) \qquad + 10\,(t+9) \quad + (u+10)$$
$$\underline{100u \qquad\qquad + 10t \qquad\quad + h}$$
$$100(h-u-1) \quad + 10(9) \qquad\; +u-h+10$$

Reversing the digits of this difference gives us:

$$100(u-h+10) + 10(9) + (h-u-1)$$

Now adding these last two expressions gives us:

$$100(9) + 10(9+9) + (10-1) = \mathbf{1{,}089}$$

It is important to stress that algebra enables us to inspect the arithmetic process, regardless of the number.

Before we leave the number 1,089, I should point out to the reader who is now so motivated to inspect this curious number further that there is still another oddity, namely, $33^2 = 1{,}089 = 65^2 - 56^2$, which is unique among two-digit numbers.

By this time you must agree that there is a particular beauty in the number **1,089**. Are you hooked?

1.12. the irrepressible number 1

There are times when we refer to beauty in nature as magical. Is magic beautiful? Some feel that when something is truly surprising and "neat" it is beautiful. From that standpoint, we will show a seemingly "magical" property in mathematics. This is one that has baffled mathematicians for many years and still no one knows why it happens. Try it; you'll like it.

We begin by asking you to follow two rules as you work with any *arbitrarily* selected number.

If the number is <u>odd</u> then multiply by 3 and add 1.
If the number is <u>even</u> then divide by 2.

Regardless of the number you select, you will always eventually end up with 1, after continued repetition of the process.

Let's try it for the *arbitrarily selected* number 12

12 is even, therefore, divide by 2 to get 6.
6 is also even so we again divide by 2 to get 3.
3 is odd, therefore, multiply by 3 and add 1 to get: $3 \bullet 3 + 1 = 10$.
10 is even, so we simply divide by 2 to get 5.
5 is odd, so we multiply by 3 and add 1 to get 16.
16 is even so we divide by 2 to get 8.
8 is even so we divide by 2 to get 4.
4 is even so we divide by 2 to get 2.
2 is even so we divide by 2 to get 1.

It is believed that no matter which number we begin with (here we started with 12) we will eventually get to 1.

This is truly remarkable! Try it for some other numbers to convince yourself that it really does work. Had we started with 17 as our arbitrarily selected number we would have required 12 steps to reach 1. Starting with 43 will require 29 steps.

Does this really work for all numbers? This is a question that has concerned mathematicians since the 1930s, and to date no answer has been found, despite monetary rewards having been offered for a proof of this conjecture. Most recently (using computers), this problem, known in the literature as the "3n + 1 Problem," has been shown to be true for the numbers up to $10^{18} - 1$.

For those who have been turned on by this curious number property, we offer you a schematic that shows

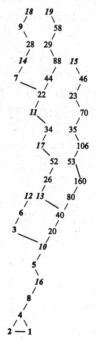

the sequence of start numbers 1–20. The numbers 1–20 can be starting points for your progression following the above rules.

Notice that you will always end up with the final loop of 4-2-1. That is, when you reach 4 you will always get to the 1 and then were you to try to continue after having arrived at the 1, you will always get back to the 1, since, by applying the rule [3 • 1 + 1 = 4], you continue in the loop: 4-2-1.

We don't want to discourage inspection of this curiosity, but we want to warn you not to get frustrated if you cannot prove that it is true in all cases, for the best mathematical minds have not been able to do this for the better part of a century!

1.13. perfect numbers

With mathematics so often touted as being the perfect science, could there be perfection within mathematics? According to tradition in number theory, we have an entity called a "perfect number." What makes a number perfect? Mathematicians just call a number "perfect" if it is **a number equal to the sum of its proper factors**. (A factor of a number is a divisor, or a number that divides the original number exactly. A proper factor is any factor besides the number itself. For example, 3 is a proper factor of 12, but 12 is an improper factor of 12.) The smallest perfect number is 6, since 6 = 1 + 2 + 3, which is the sum of all its proper factors.*

The next larger perfect number is 28, since it is equal to the sum of all of its proper factors [28 = 1 + 2 + 4 + 7 + 14]. And the next perfect number is 496 = 1 + 2 + 4 + 8 + 16 + 31 + 62 + 124 + 248, which is the sum of all the proper factors of 496.

The first four perfect numbers were known to the Greeks. They are: 6, 28, 496, and 8,128.

It was Euclid who came up with a theorem to generalize how to

---

*It is also the only number that is the sum and product of the same three numbers: 6 = 1 • 2 • 3 = 3! Also $6 = \sqrt{1^3 + 2^3 + 3^3}$. It is also cute to notice that $\frac{1}{1} = \frac{1}{2} + \frac{1}{3} + \frac{1}{6}$. By the way, while on the number 6, it is nice to realize that both 6 and its square, 36, are triangular numbers (see section 1.17).

†A prime number is one that has exactly two factors: 1 and itself. In other words it has no other divisors.

find a perfect number. He said that if $2^k - 1$ is a prime number, then $2^{k-1}(2^k - 1)$ is a perfect number, where k is a natural (or counting) number. This is to say that whenever we find a value of k that gives us a prime number† for $2^k - 1$, then we can construct a perfect number. We do not have to use all values of k, since if k is a composite number,* then $2^k - 1$ is also composite.†

Using Euclid's method for generating perfect numbers we get:

Values of k	Values of $2^{k-1}(2^k - 1)$, when $(2^k - 1)$ is a prime number
2	6
3	28
5	496
7	8,128
13	33,550,336
17	8,589,869,056
19	137,438,691,328

On observation we can notice some properties of perfect numbers. They all seem to end in either a 6 or a 28, and these are preceded by an odd digit. They also appear to be triangular numbers (see section 1.17), which are the sums of consecutive natural numbers (e.g., $496 = 1 + 2 + 3 + 4 + \cdots + 28 + 29 + 30 + 31$).

To take it a step further, every perfect number after 6 is the partial sum of the series: $1^3 + 3^3 + 5^3 + 7^3 + 9^3 + 11^3 + \cdots$. For example, $28 = 1^3 + 3^3$, and $496 = 1^3 + 3^3 + 5^3 + 7^3$.

We do not know if there are any odd perfect numbers, but none has been found yet. Using today's computers we have much greater

---

*Composite numbers are numbers that are the product of at least two factors other than 1.

†If $k = pq$, then $2^k - 1 = 2^{pq} - 1 = (2^p - 1)(2^{p(q-1)} + 2^{p(q-2)} + \cdots + 1)$. Therefore, $2^k - 1$ can only be prime when k is prime, but this does not guarantee that when k is prime, $2^k - 1$ will also be prime, as can be seen from the following values of k:

k	2	3	5	7	11	13
$2^k - 1$	3	7	31	127	2,047	8,191

where $2,047 = 23 \cdot 89$ is not a prime and so doesn't qualify.

facility at establishing more perfect numbers. If you like a challenge, you might try to find larger perfect numbers using Euclid's method.

I.I4. friendly numbers

What could possibly make two numbers friendly? Mathematicians have decided that two numbers are to be considered friendly (or as sometimes used in the more sophisticated literature, "amicable") if the sum of the proper divisors* (or factors) of one number equals the second number *and* the sum of the proper divisors of the second number equals the first number as well.

Sounds complicated? It really isn't. Just take a look at the smallest pair of friendly numbers: 220 and 284.

The proper divisors (or factors) of **220** are 1, 2, 4, 5, 10, 11, 20, 22, 44, 55, and 110. Their sum is $1 + 2 + 4 + 5 + 10 + 11 + 20 + 22 + 44 + 55 + 110 = \textbf{284}$.

The divisors of **284** are 1, 2, 4, 71, and 142, and their sum is $1 + 2 + 4 + 71 + 142 = \textbf{220}$.

This shows the two numbers can be considered *friendly numbers*.

The second pair of friendly numbers (discovered by Pierre de Fermat) is: 17,296 and 18,416.

$17{,}296 = 2^4 \bullet 23 \bullet 47$, and $18{,}416 = 2^4 \bullet 1{,}151$.

The sum of the proper factors of 17,296 is

$1 + 2 + 4 + 8 + 16 + 23 + 46 + 47 + 92 + 94 + 184 + 188 + 368 + 376 + 752 + 1{,}081 + 2{,}162 + 4324 + 8648 = \textbf{18,416}$.

---

*Proper divisors are all the divisors or factors of the number except the number itself. For example the proper divisors of 6 are 1, 2, and 3, but not 6.

The sum of the factors of 18,416 is 1 + 2 + 4 + 8 + 16 + 1,151 + 2,302 + 4,604 + 9,208 = **17,296**.

Here are a few more pairs of friendly numbers:

1,184 and 1,210
2,620 and 2,924
5,020 and 5,564
6,232 and 6,368
10,744 and 10,856
9,363,584 and 9,437,056
111,448,537,712 and 118,853,793,424

You might want to verify the above pairs' "friendliness"!

For the expert, the following is one method for finding pairs of friendly numbers:

Let $a = 3 \cdot 2^n - 1$
$b = 3 \cdot 2^{n-1} - 1$
$c = 3^2 \cdot 2^{2n-1} - 1$

where n is an integer ≥ 2, and a, b, and c are all prime numbers, then $2^n\,ab$ and $2^n\,c$ are friendly numbers.

(Notice that for $n \leq 200$, the values of $n = 2, 4$, and 7 give us prime numbers for a, b, and c.)

1.15. another friendly pair of numbers

We can always look for nice relationships between numbers. In the previous unit we experienced pairs of friendly numbers. With some creativity we can establish another form of "friendliness" between numbers. Some of them can be truly mind-boggling! Take, for example, the pair of numbers: 6,205 and 3,869.

On first look, there seems to be no apparent relationship. But with some luck and imagination we can get some fantastic results.

$$6{,}205 = 38^2 + 69^2 \text{ and } 3{,}869 = 62^2 + 05^2$$

Notice how we partitioned each number into pairs of digits and squared each pair of one number to get the other number when summing the squares.

We can even find another pair of numbers with a similar relationship. Consider these.

$$5{,}965 = 77^2 + 06^2 \text{ and } 7{,}706 = 59^2 + 65^2$$

Beyond the enjoyment of seeing this wonderful pattern, there isn't much mathematics in these examples. However, the relationship is truly amazing and worth noting. Again, mathematics has its hidden treasures.

1.16. Palindromic Numbers

Unfortunately, word puzzles and crossword puzzles are much more in vogue than mathematical puzzles. We have seen books written on palindromes, that is, words or sentences that read the same in both directions. New puzzles are always being developed with palindromes in mind. But what about palindromic numbers? They, too, can have some amusing aspects. Let us first review some palindromic words and sentences. Here are a few amusing palindromes:

RADAR
REVIVER
ROTATOR
LEPERS REPEL
MADAM I'M ADAM
STEP NOT ON PETS
NO LEMONS, NO MELON
DENNIS AND EDNA SINNED
ABLE WAS I ERE I SAW ELBA
A MAN, A PLAN, A CANAL, PANAMA
SUMS ARE NOT SET AS A TEST ON ERASMUS

Palindromic numbers are those that read the same in both directions. This leads us to consider that dates can be a source for some symmetric inspection. For example, the year 2002 is a palindrome, as is 1991.* There were several dates in October 2001 that appeared as palindromes when written in American style: 10/1/01, or 10/22/01, and others. Europeans had the ultimate palindromic moment at 8:02 P.M. on February 20, 2002, since they would have written it as 20.02, 20-02-2002. It is a bit thought provoking to come up with other palindromic dates.

Looking further, the first four powers of 11 are palindromic numbers:

$$11^1 = 11$$
$$11^2 = 121$$
$$11^3 = 1,331$$
$$11^4 = 14,641$$

A palindromic number can either be a prime number or a composite number.† For example, 151 is a prime palindrome and 171 is a composite palindrome. Yet with the exception of 11, a palindromic prime must have an odd number of digits. Try to find some other palindromic primes.

Perhaps most interesting is to see how a palindromic number can be generated from any given number. All we need to do is continually add a number to its reversal (i.e., the number written in the reverse order of digits) until a palindrome results.

For example, a palindrome can be reached with a single addition such as with the starting number 23:

23 + 32 = 55, a palindrome.

Or it might take two steps, such as with the starting number 75:

75 + 57 = 132, 132 + 231 = 363, a palindrome.

Or it might take three steps, such as with the starting number 86:

86 + 68 = 154, 154 + 451 = 605, 605 + 506 = 1,111, a palindrome.

---

*Those of us who have lived through 1991 and 2002 will be the last generation who will have lived through two palindromic years for over the next thousand years (assuming the current level of longevity).

†Integers (or positive/negative whole numbers) are either prime (no divisors other than 1 or themselves) or composite (more divisors than a prime).

The starting number 97 will require six steps to reach a palindrome, while the number 98 will require twenty-four steps.

Be cautioned about using the starting number 196; this one will go far beyond your capabilities to reach a palindrome.

There are some lovely patterns when dealing with palindromic numbers. For example, numbers that yield palindromic cubes are palindromic themselves.

You might enjoy trying to find more properties of palindromic numbers*—they're fun to play with.

1.17. fun with figurate numbers

Can numbers have a geometric shape? Although numbers do not have a geometric shape, some can be represented by dots that can be put into a regular geometric shape. Let's take a look at some of these now.

Notice how the dots form the shape of a regular polygon.

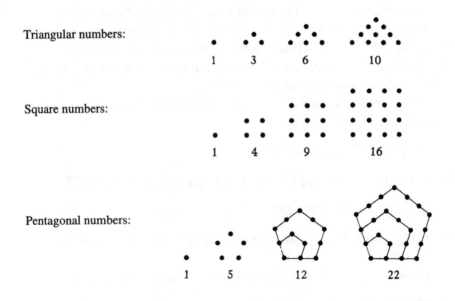

Triangular numbers:

1　3　6　10

Square numbers:

1　4　9　16

Pentagonal numbers:

1　5　12　22

---

*One source for some more information on palindromic numbers is *Teaching Secondary School Mathematics: Techniques and Enrichment Units*, 6th ed, A. S. Posamentier and J. Stepelman (Upper Saddle River, N.J.: Prentice Hall/Merrill, 2002), pp. 257–58.

Hexagonal numbers:

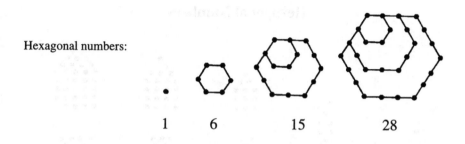

From the following arrangements of these figurate numbers you ought to be able to discover some of their properties (as we will do now). It can be fun trying to relate these numbers to one another. For example, the nth square number is equal to the sum of the nth and the $(n-1)^{th}$ triangular numbers. Another example is that the nth pentagonal number is equal to the sum of the nth square number and the $(n-1)^{th}$ triangular number. There are lots of other such relationships to be found (or discovered!).

Triangular Numbers

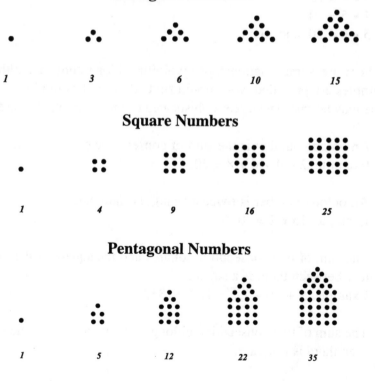

Hexagonal Numbers

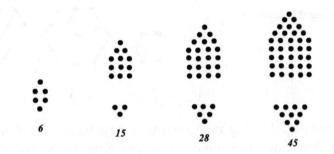

We introduce *oblong numbers*, which look like $n(n + 1)$, or rectangular arrays of dots, such as:

1 • 2 = 2
2 • 3 = 6
3 • 4 = 12
4 • 5 = 20
5 • 6 = 30, and so on.

So here are some relationships involving oblong numbers; although examples are provided, you should find additional examples to show these may be true. The more sophisticated can try to prove they are true.

An oblong number is the sum of consecutive even integers.
Example: 2 + 4 + 6 + 8 = 20.

An oblong number is twice a triangular number.
Example: 15 • 2 = 30

The sum of two consecutive squares and the square of the oblong number between them is a square.
Example: $9 + 16 + 12^2 = 169 = 13^2$

The sum of two consecutive oblong numbers and twice the square between them is a square.

Example: $12 + 20 + 2 \cdot 16 = 64 = 8^2$

The sum of an oblong number and the next square is a triangular number.
Example: $20 + 25 = 45$

The sum of a square number and the next oblong number is a triangular number.
Example: $25 + 30 = 55$

The sum of a number and the square of that number is an oblong number.
Example: $9 + 81 = 90$.

You might now wish to discover other connections between the various figurate numbers presented here.

1.18. the fabulous fibonacci numbers

There aren't many themes in mathematics that permeate more branches of mathematics than the Fibonacci numbers. They come to us from one of the most important books in Western history. This book, *Liber abaci*, written in 1202 by Leonardo Pisano, or Leonardo of Pisa, more popularly known as Fibonacci (1180–1250),* or Son of Bonacci, is the first European publication using the Hindu-Arabic numerals that are the basis for our base 10 number system. This alone would qualify it as a

---

*Fibonacci—a name he himself wrote as Filius Bonacci and later took on today's familiar name of Fibonacci—was not a clergyman, as might be expected of early scientists; rather he was a merchant who traveled extensively throughout the Islamic world and took advantage of reading all he could of the Arabic mathematical writings. He was the first to introduce the Hindu-Arabic numerals to the Christian world in his *Liber abaci* (1202 and revised in 1228), which first circulated widely in manuscript form and was first published in 1857 as *Scritti di Leonardo Pisano* (Rome: B. Buoncompagni). The book is a collection of business mathematics including linear and quadratic equations, square roots and cube roots, and other new topics, seen from the European viewpoint. He begins the book with the sentence: "These are the nine figures of the Indians 9 8 7 6 5 4 3 2 1. With these nine figures, and with the sign 0, which in Arabic is called *zephirum*, any number can be written, as will be demonstrated below." From here on he introduces the decimal position system for the first time in Europe. (Note: the word *zephirum* evolves from the Arabic word *as-sifr*, which comes from the Sanskrit word *sunya*, used in India as early as the fifth century, "sunya," referring to *empty*.)

landmark book. However, it also contains a "harmless" problem about the regeneration of rabbits. It is the solution of that problem that produces the famous Fibonacci numbers. It may be stated as follows:

> **How many pairs of rabbits will be produced in a year, beginning with a single pair, if in every month each pair bears a new pair, which becomes productive from the second month on?**

From this problem the *Fibonacci sequence* emerged. If we assume that a pair of baby (*B*) rabbits matures in one month to become offspring-producing adults (*A*), then we can set up the following chart:

Month	Pairs	No. of Pairs of Adults (*A*)	No. of Pairs of Babies (*B*)	Total Pairs
Jan. 1	A	1	0	1
Feb. 1	A B	1	1	2
Mar. 1	A B A	2	1	3
Apr. 1	A B A A B	3	2	5
May 1	A B A A B A B A	5	3	8
June 1	A B A A B A B A A B A A B	8	5	13
July 1		13	8	21
Aug. 1		21	13	34
Sept. 1		34	21	55
Oct. 1		55	34	89
Nov. 1		89	55	144
Dec. 1		144	89	233
Jan. 1		233	144	377

The number of pairs of mature rabbits living each month determines the Fibonacci sequence:

1, 1, 2, 3, 5, 8, 13, 21, 34, 55, 89, 144, 233, 377, ...

If we let f_n be the nth term of the Fibonacci sequence, then

$$f_1 = 1$$
$$f_2 = 1$$
$$f_3 = f_2 + f_1 = 1 + 1 = 2$$
$$f_4 = f_3 + f_2 = 2 + 1 = 3$$
$$f_5 = f_4 + f_3 = 3 + 2 = 5$$
$$\vdots \qquad \vdots$$
$$f_n = f_{n-1} + f_{n-2}, \text{ for an integer } n \geq 3$$

That is, each term after the first two terms is the sum of the two preceding terms.

It is normal to ask, what makes this sequence of number so spectacular? For one thing, there is a direct relationship between the Fibonacci numbers and (believe it or not) the Golden Section!* Consider quotients of successive Fibonacci numbers:

$$\frac{f_{n+1}}{f_n}$$

$$\frac{1}{1} = 1.000000000$$
$$\frac{2}{1} = 2.000000000$$
$$\frac{3}{2} = 1.500000000$$
$$\frac{5}{3} = 1.666666667$$
$$\frac{8}{5} = 1.600000000$$
$$\frac{13}{8} = 1.625000000$$
$$\frac{21}{13} = 1.615384615$$
$$\frac{34}{21} = 1.619047619$$
$$\frac{55}{34} = 1.617647059$$
$$\frac{89}{55} = 1.618218162$$
$$\frac{144}{89} = 1.617977528$$
$$\frac{233}{144} = 1.618055556$$
$$\frac{377}{233} = 1.618025751$$
$$\frac{610}{377} = 1.618037135$$
$$\frac{987}{610} = 1.618032787$$

*The Golden Section refers to one of the most unusual ratios in mathematics. There will be more on the Golden Section later in the book.

Furthermore, you can see the unit on "Some Algebra on the Golden Section" (see section 4.18) to notice that successive powers of ϕ* present us with the Fibonacci numbers.

$$\phi^2 = \phi + 1$$
$$\phi^3 = 2\phi + 1$$
$$\phi^4 = 3\phi + 2$$
$$\phi^5 = 5\phi + 3$$
$$\phi^6 = 8\phi + 5$$
$$\phi^7 = 13\phi + 8$$

The connection should be obvious. Just look at the coefficients and the constants. There you have the Fibonacci numbers. This is quite incredible; two completely (seemingly) unrelated things suddenly in close relationship to one another. That's what makes mathematics so wonderful!

1.19. getting into an endless loop

This unit demonstrates an unusual phenomenon that arises out of our decimal number system. There isn't much you can do with it, other than to marvel at the outcome. This is not something we can prove true for all cases, yet no numbers have been found for which it won't work. That, in itself, suffices to establish that it is apparently always true. You may wish to use a calculator, unless you enjoy subtraction.

Follow the steps below:

1. Begin by selecting a four-digit number (except one that has all the same digits).
2. Rearrange the digits of the number so that they form the largest number possible. (That means write the number with the digits in descending order.)
3. Then rearrange the digits of the number so that they form the smallest number possible. (That means write the number with the digits in ascending order.)

---

*ϕ represents the Golden Ratio.

4. Subtract these two numbers (obviously, the smaller from the larger).
5. Take this difference and continue the process, over and over and over, until you notice something disturbing happening. Don't give up before something unusual happens.

Eventually you will arrive at the number 6,174, perhaps after one subtraction, or after several subtractions. When you do, you will find yourself in an endless loop.

When you have reached the loop, remember that you began with an arbitrarily selected number. Isn't this quite an astonishing result? Some readers might be motivated to investigate this further. Others will just sit back in awe. Either way you've been charmed again with the beauty of mathematics.

Here is an example of how this works with our arbitrarily selected starting number 3,203.

The largest number formed with these digits is: 3,320
The smallest number formed with these digits is: 0,233
The difference is: 3,087

Now using this number, 3,087 we continue the process.

The largest number formed with these digits is: 8,730
The smallest number formed with these digits is: 0,378
The difference is: 8,352

Again, we repeat the process.

The largest number formed with these digits is: 8,532
The smallest number formed with these digits is: 2,358
The difference is: **6,174**

The largest number formed with these digits is: 7,641
The smallest number formed with these digits is: 1,467
The difference is: **6,174**

And so the loop is formed, since you keep on getting 6,174. Remember, all this began with an arbitrarily selected number and will always end up with the number 6,174, which then gets you into an endless loop (i.e., we continuously get back to 6,174).

1.20. a power loop

Can you imagine that a number is equal to the sum of the cubes of its digits? This is true for only five numbers. Here are these five most unusual numbers.

$$1 \rightarrow 1^3 = 1$$
$$153 \rightarrow 1^3 + 5^3 + 3^3 = 1 + 125 + 27 = 153$$
$$370 \rightarrow 3^3 + 7^3 + 0^3 = 27 + 343 + 0 = 370$$
$$371 \rightarrow 3^3 + 7^3 + 1^3 = 27 + 343 + 1 = 371$$
$$407 \rightarrow 4^3 + 0^3 + 7^3 = 64 + 0 + 343 = 407$$

Take a moment to appreciate these spectacular results and take note that these are the *only* such numbers for which this is true.

Taking sums of the powers of the digits of a number leads to interesting results. We can extend this procedure to get a lovely (and not to mention, surprising) technique that can help refamiliarize you with the powers of numbers and at the same time reach a startling conclusion.

Select any number and then find the sum of the cubes of the digits, just as we did above. Of course, for any number other than those above, you will have reached a number different from the one with which you started. Repeat this process with each succeeding sum until you get into a "loop." A loop can be easily recognized when you reach a number that you already reached earlier. This should become clearer with an example.

Let's begin with the (arbitrarily selected) number **352** and find the sum of the cubes of the digits.

The sum of the cubes of the digits of 352 is: $3^3 + 5^3 + 2^3 = 27 + 125 + 8 = 160$

Now we use this sum, 160, and repeat the process:

The sum of the cubes of the digits of 160 is: $1^3 + 6^3 + 0^3 = 1 + 216 + 0 = 217.$

Repeat the process with 217:

The sum of the cubes of the digits of 217 is: $2^3 + 1^3 + 7^3 = 8 + 1 + 343 = 352$

Surprise! This is the same number (352) we started with.

You might think it would have been easier to begin by taking squares. You are in for a surprise. Let's try this with the number 123.

Beginning with 123, the sum of the squares of the digits is: $1^2 + 2^2 + 3^2 = 1 + 4 + 9 = 14.$

1. Now using 14, the sum of the squares of the digits is: $1^2 + 4^2 = 1 + 16 = 17.$
2. Now using 17, the sum of the squares of the digits is: $1^2 + 7^2 = 1 + 49 = 50.$
3. Now using 50, the sum of the squares of the digits is: $5^2 + 0^2 = 25.$
4. Now using 25, the sum of the squares of the digits is: $2^2 + 5^2 = 4 + 25 = 29.$
5. Now using 29, the sum of the squares of the digits is: $2^2 + 9^2 = 85.$
6. Now using 85, the sum of the squares of the digits is: $8^2 + 5^2 = 64 + 25 = 89.$
7. Now using 89, the sum of the squares of the digits is: $8^2 + 9^2 = 64 + 81 = 145.$
8. Now using 145, the sum of the squares of the digits is: $1^2 + 4^2 + 5^2 = 1 + 16 + 25 = 42.$

9. Now using 42, the sum of the squares of the digits is: $4^2 + 2^2 = 16 + 4 = 20$.
10. Now using 20, the sum of the squares of the digits is: $2^2 + 0^2 = 4 + 0 = 4$.
11. Now using 4, the square of the digit is: $4^2 = 16$.
12. Now using 16, the sum of the squares of the digits is: $1^2 + 6^2 = 1 + 36 = 37$.
13. Now using 37, the sum of the squares of the digits is: $3^2 + 7^2 = 9 + 49 = 58$.
14. Now using 58, the sum of the squares of the digits is: $5^2 + 8^2 = 25 + 64 = 89$.

Notice that the last sum, 89, is the same as in step 6, and so a repetition will now begin after step 14. This indicates that we would continue in a loop.

You may want to experiment with the sums of the powers of the digits of any number and see what interesting results it may lead to. Look for patterns of loops, and perhaps determine the extent of a loop based on the nature of the original number.

In any case, this intriguing charmer can be fun just as it is presented here or it can be a source for further investigation by interested readers.

1.21. a factorial loop

This little charming unit will show an unusual relationship for certain numbers. But before we begin, we shall review the factorial notation. The definition of $n!$ is $n! = 1 \bullet 2 \bullet 3 \bullet 4 \cdots (n-1) \bullet n.$*

Now that you have an understanding of the factorial concept, find the sum of the factorials of the digits of 145.

$1! + 4! + 5! = 1 + 24 + 120 = 145$.

Surprise! We're back to 145.

---

*The factorial notation "!" represents the product of all the integers equal to and less than the number with the factorial notation. For example, $5! = (5)(4)(3)(2)(1) = 120$, $8! = (8)(7)(6)(5)(4)(3)(2)(1) = 40,320$, and $n! = (n)(n-1)(n-2)(n-3) \ldots (3)(2)(1)$. Notice here we are using the parentheses to indicate multiplication. By the way $0! = 1$, by definition (i.e., mathematicians simply agree to this value since it keeps things consistent).

Only for certain numbers will the sum of the factorials of the digits equal the number itself.

Let's try this again with the number 40,585.

That is, $4! + 0! + 5! + 8! + 5! = 24 + 1 + 120 + 40,320 + 120 = 40,585$

It is natural that you would expect this to be true for just about any number. Well, just try another number. Chances are that it will not work. Suppose we begin with the number 871.

Using this procedure you will get: $8! + 7! + 1! = 40,320 + 5,040 + 1 = 45,361$. At this point you may feel that you have failed. Not so fast. Try this procedure again with the previous result 45,361.

This yields: $4! + 5! + 3! + 6! + 1! = 24 + 120 + 6 + 720 + 1 = 871$. Isn't this the very number we started with? Again we formed a loop.

If you repeat this with the number 872, you will get $8! + 7! + 2! = 40,320 + 5,040 + 2 = 45,362$. Then repeating the process will give you: $4! + 5! + 3! + 6! + 2! = 24 + 120 + 6 + 720 + 2 = 872$. Again we're in a loop.

Some people are quick to form generalizations, so they might conclude that if the scheme of summing factorials of the digits of a number doesn't get you back to the original number, then try it again and it ought to work. Of course you can "stack the deck" by considering the number 169. Two cycles do not seem to present a loop. So proceed through one more cycle. And sure enough, the third cycle leads back to the original number.

Starting number	Sum of the factorials
169	$1! + 6! + 9! = 363,601$
363,601	$3! + 6! + 3! + 6! + 0! + 1! = 6 + 720 + 6 = 720 + 1 + 1 = 1,454$
1,454	$1! + 4! + 5! + 4! = 1 + 24 + 120 + 24 = 169$

Be careful about drawing conclusions. These factorial oddities are not so pervasive that you should try to find others. There are "within reach" three groups of such loops. We can organize them according to the number of times you have to repeat the process to reach the original number. We will call these repetitions "cycles."

Here is a summary of the way our numbers behave in this factorial loop.

1 cycle	1; 2; 145; 40,585
2 cycles	871; 45,361 and 872; 45,362
3 cycles	169; 363,601; 1,454

The factorial loops shown in this charming little number oddity can be fun, but you ought to be cautioned that there are no other such numbers less than 2,000,000 for which this works. So don't waste your time in a fruitless search. Just appreciate the few little beauties!

1.22. the irrationality of $\sqrt{2}$

When we say that the $\sqrt{2}$ is irrational, what does that mean? It may be wise to inspect the word *irrational* with regard to its meaning in English.

Irrational means not rational.

Not rational means it cannot be expressed as a ratio of two integers.

Not expressible as a ratio means it cannot be expressed as a common fraction.

That is, there is no fraction $\frac{a}{b} = \sqrt{2}$, where a and b are integers.*

If we compute with a calculator we will get:

$\sqrt{2}$ = 1.4142135623730950488016887242096980785696718753 76948 07317667973799073247846210703885038753432764157 2...

---

*Integers, you will recall, are essentially whole numbers and can be positive, negative, or zero.

Notice that there is no pattern among the digits, and there is no repetition even of groups of digits. Does this mean that all rational fractions will have a period of digits? Let's inspect a few common fractions.

$\frac{1}{7}$ = 0.142857<u>142857</u>142857<u>142857</u> ... , which can be written as: 0.$\overline{142857}$ (a six-digit period)*

Suppose we consider the fraction $\frac{1}{109}$:

$\frac{1}{109}$ = 0.0091743119266055045871559633027522935779816513761 46788990825688073394495412844036697247706422018348 6...

Here we have calculated its value to 100 places and no period appears. Does this mean that the fraction is irrational? This would destroy our nice definition above. We can try to calculate the value a bit more accurately, that is, say, to another 12 places.

$\frac{1}{109}$ = 0.0091743119266055045871559633027522935779816513761 46788990825688073394495412844036697247706422018348 62385 3211<u>0091</u>

Suddenly it looks as though a pattern may be appearing; the 0091 also began the period.

We carry out our calculation further to 220 places and notice that a 108-digit period emerges.

$\frac{1}{109}$ = 0.0091743119266055045871559633027522935779816513761 46788990825688073394495412844036697247706422018348 62385 3211<u>009174311926605504587155963302752293577981651376146 7889908256880733944954128440366972477064220183486238532 11</u> 009174

If we carry out the calculation to 330 places, the pattern becomes clearer.

_____

*A period of digits is a group of digits that repeats, as in the case of this example.

$\frac{1}{109}$ = 0.**0091743119266055045871559633027522935779816513761**
46788990825688073394495412844036697247706422018348623 85
321100917431192660550458715596330275229357798165137 61467
88990825688073394495412844036697247706422018348623853211
00917431192660550458715596330275229357798165137614678899
082568807339449541284403669724770642201834862385321100091
74

We might be able to conclude (albeit without proof) that a common fraction results in a decimal equivalent that has a repeating period of digits. Some common ones we already are familiar with, such as:

$$\frac{1}{3} = .333333333\overline{3}$$
$$\frac{1}{13} = 0.0769230769230769230769230\overline{769230}$$

To this point, we saw that a common fraction will result in a repeating decimal, sometimes with a very long period (e.g., $\frac{1}{109}$) and sometimes with a very short period (e.g., $\frac{1}{3}$). It would appear, from the rather flimsy evidence so far, that a fraction results in a repeating decimal and an irrational number does not. Yet this does not prove that an irrational number cannot be expressed as a fraction.

Here is a cute proof for the more ambitious (or curious) reader that $\sqrt{2}$ cannot be expressed as a common fraction and, therefore by definition, is irrational.

Suppose $\frac{a}{b}$ is a fraction in lowest terms, which means that a and b do not have a common factor.

We shall prove this by a method of contradiction. That is, we will start with a premise, which we will then show is false, resulting in a reversal of the premise.

Suppose $\frac{a}{b} = \sqrt{2}$. Then $\frac{a^2}{b^2} = 2$, or $a^2 = 2b^2$, which implies that a^2 and a are divisible by 2; written another way: $a = 2r$,* where r is an integer.
Then $4r^2 = 2b^2$, or $2r^2 = b^2$.

---

*Any even number can be expressed, or written, as 2 times some other integer. In this case we used the integer r.

So we have b^2 or b is divisible by 2.

This contradicts the beginning assumption about the fact that a and b have no common factor, so $\sqrt{2}$ cannot be expressed as a common fraction.

Understanding this proof may be a bit strenuous for some, but a slow and careful step-by-step consideration of it should make it understandable for most readers.

I.23. Sums of Consecutive Integers

Which numbers can be expressed as the sum of consecutive integers? You may want to experiment a bit before trying to generate a rule. Try to express the first batch of natural numbers as the sum of consecutive integers. We will provide some in the following list.

2 = not possible	21 = 1 + 2 + 3 + 4 + 5 + 6
3 = 1 + 2	22 = 4 + 5 + 6 + 7
4 = not possible	23 = 11 + 12
5 = 2 + 3	24 = 7 + 8 + 9
6 = 1 + 2 + 3	25 = 12 + 13
7 = 3 + 4	26 = 5 + 6 + 7 + 8
8 = not possible	27 = 8 + 9 + 10
9 = 4 + 5	28 = 1 + 2 + 3 + 4 + 5 + 6 + 7
10 = 1 + 2 + 3 + 4	29 = 14 + 15
11 = 5 + 6	30 = 4 + 5 + 6 + 7 + 8
12 = 3 + 4 + 5	31 = 15 + 16
13 = 6 + 7	32 = not possible
14 = 2 + 3 + 4 + 5	33 = 10 + 11 + 12
15 = 4 + 5 + 6	34 = 7 + 8 + 9 + 10
16 = not possible	35 = 17 + 18
17 = 8 + 9	36 = 1 + 2 + 3 + 4 + 5 + 6 + 7 + 8
18 = 5 + 6 + 7	37 = 18 + 19
19 = 9 + 10	38 = 8 + 9 + 10 + 11
20 = 2 + 3 + 4 + 5 + 6	39 = 19 + 20
	40 = 6 + 7 + 8 + 9 + 10

These consecutive number sum representations are clearly not unique. For example, 30 can be expressed in more than one way:

30 = 9 + 10 + 11, or 30 = 6 + 7 + 8 + 9.

An inspection of the table shows that those where a consecutive number sum was not possible were powers of 2.

This is an interesting fact. It is not something that one would expect. By making a list of these consecutive number sums, you ought to begin to see patterns. Clearly the triangular numbers are equal to the sum of the first n natural numbers.* A multiple of 3, say $3n$, can always be represented by the sum: $(n-1) + n + (n+1)$. You may discover other patterns. That's part of the fun of it (not to mention its instructional value—seeing number patterns and relationships).

For the purists and the more ambitious reader or one who is properly math-prepared, we now prove this (until-now) conjecture. First we will establish when a number can be expressed as a sum of at least two consecutive positive integers.

Let us analyze what values can be taken by the sum (S) of (two or more) consecutive positive integers from a to b. ($b > a$)

$$S = a + (a+1) + (a+2) + \ldots + (b-1) + b = \left(\tfrac{a+b}{2}\right)(b - a + 1)$$

by applying the formula for the sum of an arithmetic series.† Then doubling both sides we get: $2S = (a+b)(b-a+1)$

Calling $(a+b) = x$ and $(b - a + 1) = y$, we can note that x and y are both integers and that since their sum, $x + y = 2b + 1$, is odd, one of x, y is odd and the other is even. Note that $2S = xy$.

Case 1. S is a power of 2.

Let $S = 2^n$. We have $2(2^n) = xy$, or $2^{n+1} = xy$. The only way we can express 2^{n+1} as a product of an even and an odd number is if the odd number is 1. If $x = a + b = 1$, then a and b cannot be positive integers. If $y = b - a + 1 = 1$ then we have $a = b$ which also cannot occur. Therefore, S cannot be a power of 2.

---

*Remember that the natural numbers are the counting numbers: 1, 2, 3, 4, 5, 6, ...

†The sum of an arithmetic series is $S = \tfrac{n}{2}(a + l)$, where n is the number of terms, and a is the first term and l is the last term.

Case 2. S is not a power of 2.

Let $S = m2^n$, where m is an odd number greater than 1. We have $2(m2^n) = xy$, or $m2^{(n+1)} = xy$. We will now find positive integers a and b such that $b > a$ and $S = a + (a+1) + \ldots + b$.

The two numbers 2^{n+1} and m are not equal, since one is odd and the other is even. Therefore one is bigger than the other. Assign x to be the bigger one and y to be the smaller one. This assignment gives us a solution for a and b, as $x + y = 2b + 1$, giving a positive integer value for b, and $x - y = 2a - 1$, giving a positive integer value for a. Also, $y = b - a + 1 > 1$, so $b > a$, as required. We have obtained a and b.

Therefore, for any S that is not a power of 2 we can find positive integers a and b, $b > a$, such that $S = a + (a+1) + \ldots + b$.

Just to recap, a number can be expressed as a sum of (at least two) consecutive positive integers if and only if the number is not a power of 2. This may appear harmless, but you should be aware that you just accomplished some nice mathematics, while admiring the beauty of the subject.

2

Some
Arithmetic
Marvels

now that the calculator is so ubiquitous, we do not have to see arithmetic as a burden anymore. We used to have to memorize the arithmetic algorithms* and didn't have much chance to enjoy the nature and behavior of the arithmetic. There are clever shortcuts around some arithmetic procedures and there are "tricks" for avoiding cumbersome arithmetic processes. For example, an almost visual inspection of a number to determine its divisors can be a very useful technique. Additionally, some alternate forms of multiplication are more amusing than useful. In either case, they help to bring the topic of arithmetic to life, and provide some refreshing insights into the subject.

---

*An algorithm is a step-by-step problem-solving procedure, especially an established, recursive computational procedure for solving a problem in a finite number of steps. Examples of arithmetic algorithms are the procedures we use to multiply or divide numbers.

This chapter also includes some recreational units that should strengthen your understanding of the nature of arithmetic processes. For example, the unit on "alphametics" provides an opportunity to really work within the place value system, beyond the mere rote algorithms. You will truly have fun with this unit, as you will be trying to apply your mature understanding to an elementary situation.

The unit on the Rule of 72 is of particular use when dealing with compound interest and wanting to get some insight into the power of the compounding effect. The level at which this unit could be tackled depends on your motivation. It can be considered as an algorithm or inspected closely to discover why it works as it does.

Essentially, this chapter presents a variety of aspects of arithmetic applications, with the sole purpose of turning on the reader to a subject that has been falsely labeled tedious.

2.1. multiplying by 11

The simpler a mathematical "trick" is, the more attractive it tends to be. Here is a very nifty way to multiply by 11. This one always gets a rise out of the unsuspecting mathematics-phobic person, because it is so simple that we do not even need a calculator!

The rule is very simple: *To multiply a two-digit number by 11 just add the two digits and place this sum between the two digits (with appropriate "carries" as explained below).*

Let's try using this technique. Suppose you need to multiply 45 by 11. According to the rule, add 4 and 5 and place the resulting sum between the 4 and 5 to get 495.

This can get a bit more difficult. Suppose the sum of the two digits you added results in a two-digit number. What do we do in a case like that? We no longer have a single digit to place between the two original digits. So if the sum of the two digits is greater than 9, then we place the units digit between the two digits of the number being multiplied by 11 and "carry" the tens digit to be added to the hundreds digit of the multiplicand.* Let's try it with 78 • 11.

7 + 8 = 15. We place the 5 between the 7 and 8, and add the 1 to the 7, to get [7 + 1][5][8] or 858.

You may legitimately ask if the rule also holds when 11 is multiplied by a number of more than two digits.

Let's go right for a larger number such as 12,345 and multiply it by 11.

Here we begin at the right-side digit and add every pair of digits going right to left, placing the sums in order between the end digits of the number.

1[1 + 2][2 + 3][3 + 4][4 + 5]5 = 135,795.

If the sum of two digits is greater than 9, then use the procedure described before: place the units digit appropriately and carry the tens digit for each paired sum to the next pair, right to left. We will do one of these here.

Multiply 6,789 by 11.

We carry the process step by step (numbers in italics are the numbers carried):
Replacement for page 74 to be placed at the bottom of page 73.

6[6 + 7][7 + 8][8 + 9]9	Add each pair of digits between the end digits.
6[6 + 7][7 + 8][17]9	Add 8 + 9 = 17.
6[6 + 7][7 + 8 + *1*][7]9	Carry the 1 (from the 17) to the next sum.
6[6 + 7][16][7]9	Add 7 + 8 + 1 = 16.
6[6 + 7 + *1*][6][7]9	Carry the 1 (from the 16) to the next sum.
6[14][6][7]9	Add 6 + 7 + 1 = 14.
[6 + *1*][4][6][7]9	Carry the 1 (from the 14) to the next sum.
[7][4][6][7]9	Add 6 + 1 = 7, and so the product is 74,679.

---

*The multiplicand is the number that is multiplied by another number, the multiplier. In arithmetic, the multiplicand and the multiplier are interchangeable, depending on how the problem is stated, because the result is the same if the two are reversed. For example, 2 × 3 and 3 × 2.

This rule for multiplying by 11 ought to be shared with your friends. Not only will they be impressed with your cleverness, they may also appreciate knowing this shortcut.

2.2. When is a number divisible by 11?

At the oddest times the issue can come up as to whether a number is divisible by 11. If you have a calculator on hand, the problem is easily solved. But that is not always the case. Besides, there is such a clever "rule" for testing for divisibility by 11 that it is worth knowing just for its charm.

The rule is quite simple: *If the difference of the sums of the alternate digits is divisible by 11, then the original number is also divisible by 11.*

Sounds a bit complicated, but it really isn't. Let us take this rule a piece at a time. The sums of the alternate digits means you begin at one end of the number taking the first, third, fifth, etc. digits and add them. Then add the remaining (even placed) digits. Subtract the two sums and inspect for divisibility by 11.

It is probably best demonstrated through an example. We shall test 768,614 for divisibility by 11.
Sums of the alternate digits are: $7 + 8 + 1 = 16$, and $6 + 6 + 4 = 16$. The difference of these two sums, $16 - 16 = 0$, which is divisible by 11.* Therefore we can conclude that 768,614 is divisible by 11.

Another example might be helpful to firm up your understanding of this procedure. To determine if 918,082 is divisible by 11, find the sums of the alternate digits:

$9 + 8 + 8 = 25$, and $1 + 0 + 2 = 3.$

---

*Remember $\frac{0}{11} = 0$. However, $\frac{11}{0} =$ undefined.

Their difference is 25 − 3 = 22, which is divisible by 11, and so the number 918,082 is divisible by 11.*

Now just practice with this rule. You'll find it not only very helpful, but also an expression of the power and consistency of mathematics.

2.3. When is a number divisible by 3 or 9?

There are moments in everyday situations where the nature of a number can be useful to know, especially if it can be done instantly "in your head." For example, a restaurant bill of $71.22 needs to be split into three equal parts. Before actually doing the division, the thought about whether or not it is possible to split the bill equally may come into question—only mentally, of course! Wouldn't it be nice if there were some mental arithmetic shortcut for determining this? Well, here comes mathematics to the rescue. I am going to provide you with a rule to determine if a number is divisible by 3 and (as an extra bonus) divisible by 9.

The rule, simply stated, is: *If the sum of the digits of a number is divisible by 3 (or 9), then the original number is divisible by 3 (or 9).*†

Perhaps an example would best firm up an understanding of this rule. Consider the number 296,357. Let's test it for divisibility by 3 (or 9).

The sum of the digits is 2 + 9 + 6 + 3 + 5 + 7 = 32, which is not divisible by 3 or 9.

---

*For the interested reader, here is a brief discussion about why this rule works as it does. Consider the base-10 number ab,cde, whose value can be expressed as

$N = 10^4a + 10^3b + 10^2c + 10d + e = (11-1)^4a + (11-1)^3b + (11-1)^2c + (11-1)d + e$

$= [11M + (-1)^4]a + [11M + (-1)^3]b + [11M + (-1)^2]c + [11 + (-1)]d + e$

$= 11M [a + b + c + d] + a - b + c - d + e$, which implies that divisibility by 11 of N depends on the divisibility of: $a - b + c - d + e$, which implies that divisibility by 11 of N depends on the divisibility by 11 of: $a - b + c - d + e = (a + c + e) - (b + d)$, which is the difference of the sums of the alternate digits.

Note: $11M$ refers to a multiple of 11.

†For the interested reader, here is a brief discussion about why this rule works as it does. Consider the base-10 number ab,cde, whose value can be expressed as

$N = 10^4a + 10^3b + 10^2c + 10d + e = (9+1)^4a + (9+1)^3b + (9+1)^2c + (9+1)d + e$

$= [9M + (1)^4]a + [9M + (1)^3]b + [9M + (1)^2]c + [9 + (1)]d + e$

$= 9M [a + b + c + d] + a + b + c + d + e$, which implies that divisibility by 9 of N depends on the divisibility of: $a + b + c + d + e$, which is the sum of the digits.

Note: $9M$ refers to a multiple of 9.

Therefore the original number is neither divisible by 3 nor 9.

Another example: Is the number 457,875 divisible by 3 or 9? The sum of the digits is $4 + 5 + 7 + 8 + 7 + 5 = 36$,* which is divisible by 9 (and then, of course, by 3 as well), so the number 457,875 is divisible by 3 and by 9.

A last example: Is the number 27,987 divisible by 3 or 9? The sum of the digits is $2 + 7 + 9 + 8 + 7 = 33$, which is divisible by 3 but not by 9; therefore, the number 27,987 is divisible by 3 and not by 9.

Now that you are an expert at determining if a number is divisible by 3 or 9, we can go back to the original question about the divisibility of the restaurant bill of $71.22. Can it be divided into three equal parts? Because $7 + 1 + 2 + 2 = 12$, and 12 is divisible by three, then $71.22 is divisible by 3.

2.4. divisibility by prime numbers

In the previous unit, we presented a nifty little technique for determining if a number is divisible by 3 or by 9. Most adults can determine when a number is divisible by 2 or by 5, simply by looking at the last digit (i.e., the units digit) of the number. That is, if the last digit is an even number (such as 2, 4, 6, 8, 0), then the number will be divisible by 2.† Similarly for 5: If the last digit of the number being inspected for divisibility is either a 0 or 5, then the number itself will be divisible by 5.‡ The question then is: are there also rules for divisibility by other numbers? What about prime numbers?

With the proliferation of the calculator there is no longer a crying need to be able to detect by which numbers a given number is divisible. You can simply do the division on a calculator. Yet, for a better

---

*If by some remote chance it is not immediately clear to you if the sum of the digits is divisible by 3 or 9, then take the sum of the digits of this resulting number and continue the process until you can visually make a determination of divisibility by 3 or 9.

†Incidentally, if the number formed by the last two digits is divisible by 4, then the number itself is divisible by 4. Also, if the number formed by the last three digits is divisible by 8, then the number itself is divisible by 8. You ought to be able to extend this rule to divisibility by higher powers of 2 as well.

‡If the number formed by the last two digits is divisible by 25, then the number itself is divisible by 25. This is analogous to the rule for powers of 2. Have you guessed what the relationship here is between powers of 2 and 5? Yes, they are the factors of 10, the basis of our decimal number system.

appreciation of mathematics, divisibility rules provide an interesting "window" into the nature of numbers and their properties. For this reason (among others), the topic of divisibility still finds a place on the mathematics-learning spectrum.

It has always been most perplexing to establish rules for divisibility by prime numbers. This is especially true for the rule for divisibility by 7, which follows a series of very nifty divisibility rules for the numbers 2 through 6.* As you will soon see, some of the divisibility rules for prime numbers are almost as cumbersome as an actual division algorithm, yet they are fun, and, believe it or not, can come in handy.

Let us consider the rule for divisibility by 7 and then, as we inspect it, see how this can be generalized for other prime numbers.

The rule for divisibility by 7: *Delete the last digit from the given number, and then subtract twice this deleted digit from the remaining number. If the result is divisible by 7, the original number is divisible by 7. This process may be repeated if the result is too large for simple visual inspection of divisibility by 7.*

Let's try an example to see how this rule works. Suppose we want to test the number 876,547 for divisibility by 7.

Begin with 876,547 and delete its units digit, 7, and then subtract its double, 14, from the remaining number: $87,654 - 14 = 87,640$.

Since we cannot yet visually inspect the resulting number for divisibility by 7, we continue the process.

Continue with the resulting number 87,640 and delete its units digit, 0, and subtract its double, still 0, from the remaining number; we get: $8,764 - 0 = 8,764$.

Since this did not change the resulting number, 8,764, as we seek to check for divisibility by 7, we continue the process.

Continue with the resulting number 8,764 and delete its units digit, 4, and subtract its double, 8, from the remaining number; we get: $876 - 8 = 868$.

Since we still cannot visually inspect the resulting number, 868, for divisibility by 7, we continue the process.

---

*The rule for divisibility by 6 is simply to apply the rules for divisibility by 2 and by 3—both must hold true for a number to be divisible by 6.

Continue with the resulting number 868 and delete its units digit, 8, and subtract its double, 16, from the remaining number; we get: $86 - 16 = 70$, which is divisible by 7. Therefore the number 876,547 *is* divisible by 7.

Before continuing with our discussion of divisibility of prime numbers you ought to practice this rule with a few randomly selected numbers, and then check your results with a calculator.

Now for the beauty of mathematics! Why does this rather strange procedure actually work? To see why it works is the wonderful thing about mathematics. It doesn't do things that for the most part we cannot justify.* This will all make sense to you after you see what is happening with this procedure.

To justify the technique of determining divisibility by 7, consider the various possible terminal digits (that you are "dropping") and the corresponding subtraction that is actually being done by dropping the last digit. In the chart below you will see how dropping the terminal digit and doubling it to get the units digit of the number being subtracted gives us in each case a multiple of 7. That is, you have taken "bundles of 7" away from the original number. Therefore, if the remaining number is divisible by 7, then so is the original number, because you have separated the original number into two parts, each of which is divisible by 7, and therefore, the entire number must be divisible by 7.

Terminal digit	Number subtracted from original	Terminal digit	Number subtracted from original
1	$20 + 1 = 21 = 3 \bullet 7$	5	$100 + 5 = 105 = 15 \bullet 7$
2	$40 + 2 = 42 = 6 \bullet 7$	6	$120 + 6 = 126 = 18 \bullet 7$
3	$60 + 3 = 63 = 9 \bullet 7$	7	$140 + 7 = 147 = 21 \bullet 7$
4	$80 + 4 = 84 = 12 \bullet 7$	8	$160 + 8 = 168 = 24 \bullet 7$
		9	$180 + 9 = 189 = 27 \bullet 7$

The rule for divisibility by 13: *This is similar to the rule for testing divisibility by 7, except that the 7 is replaced by 13 and instead of sub-*

---

*There are a few phenomena in mathematics that have not yet found an acceptable justification (or proof), but that doesn't mean we won't find one in the future. It took us 350 years to justify Fermat's conjecture! It was done by Dr. Andrew Wiles a few years ago.

tracting twice the deleted digit, we subtract nine times the deleted digit each time.

Let's check for divisibility by 13 for the number 5,616.

Begin with 5,616 and delete its units digit, 6, and subtract 54, which is 6 • 9, from the remaining number:

$$561 - 54 = 507$$

Since we still cannot visually inspect the resulting number for divisibility by 13, we continue the process.

Continue with the resulting number 507 and delete its units digit (7) and subtract nine times this digit ($9 • 7 = 63$) from the remaining number:

$50 - 63 = -13$, which is divisible by 13, and therefore, the original number is divisible by 13.

To determine the "multiplier," 9, we sought the smallest multiple of 13 that ends in a 1. That was 91, where the tens digit is 9 times the units digit. Once again consider the various possible terminal digits and the corresponding subtractions in the following table.

Terminal digit	Number subtracted from original	Terminal digit	Number subtracted from original
1	$90 + 1 = 91 = 7 • 13$	5	$450 + 5 = 455 = 35 • 13$
2	$180 + 2 = 182 = 14 • 13$	6	$540 + 6 = 546 = 42 • 13$
3	$270 + 3 = 273 = 21 • 13$	7	$630 + 7 = 637 = 49 • 13$
4	$360 + 4 = 364 = 28 • 13$	8	$720 + 8 = 728 = 56 • 13$
		9	$810 + 9 = 819 = 63 • 13$

In each case a multiple of 13 is being subtracted one or more times from the original number. Hence, if the remaining number is divisible by 13, then the original number is divisible by 13.

Divisibility by 17: *Delete the units digit and subtract five times the deleted digit each time from the remaining number until you reach a number small enough to determine its divisibility by 17.*

We justify the rule for divisibility by 17 as we did the rules for 7 and 13. Each step of the procedure subtracts a "bunch of 17s" from the original number until we reduce the number to a manageable size and can make a visual inspection of divisibility by 17.

The patterns developed in the preceding three divisibility rules (for 7, 13, and 17) should lead you to develop similar rules for testing divisibility by larger primes. The following chart presents the "multipliers" of the deleted digits for various primes.

To test divisibility by	7	11	13	17	19*	23	29	31	37	41	43	47
Multiplier	2	1	9	5	17	16	26	3	11	4	30	14

You may want to extend this chart. It's fun, and it will increase your perception of mathematics. You may also want to extend your knowledge of divisibility rules to include composite (i.e., non-prime) numbers. Why the following rule refers to relatively prime factors and not just any factors is something that will sharpen your understanding of number properties. Perhaps the easiest response to this question is that relatively prime factors have independent divisibility rules, whereas other factors may not.

Divisibility by composite numbers: *A given number is divisible by a composite number if it is divisible by each of its relatively prime† factors.*

The chart below offers illustrations of this rule. You should complete the chart to 48.

To be divisible by	6	10	12	15	18	21	24	26	28
The number must be divisible by	2,3	2,5	3,4	3,5	2,9	3,7	3,8	2,13	4,7

At this juncture you have not only a rather comprehensive list of rules for testing divisibility, but also an interesting insight into elementary

---

*There is another curious rule for divisibility by 19; that is, delete the last digit of the number being tested for divisibility by 19 and *add* its double to the remaining number. Continue this process until you can recognize divisibility by 19.

†Two numbers are relatively prime if they have no common factors other than 1.

number theory. Practice using these rules (to instill greater familiarity) and try to develop rules to test divisibility by other numbers in base 10 and to generalize these rules to other bases. Unfortunately, lack of space prevents a more detailed development here. Yet, I hope to have now whet your appetite.

2.5. the russian peasant's method of multiplication

It is said that the Russian peasants used a rather strange, perhaps even primitive, method to multiply two numbers. It is actually quite simple, yet somewhat cumbersome. Let's take a look at it.

Consider the problem of finding the product of 43 • 92:

Let's work this multiplication together. We begin by setting up a chart of two columns with the two members of the product in the first row. Below you will see the 43 and 92 heading up the columns. One column will be formed by doubling each number to get the next, while the other column will be formed by taking half the number and dropping the remainder. For convenience, our first column will be the doubling column, and the second column will be the halving column. Notice that by halving the odd number such as 23 (the third number in the second column) we get 11 with a remainder of 1 and we simply drop the 1. The rest of this halving process should now be clear. The process ends when the number in the "halving" column is 1.

doubling	halving
43	92◊
86	46◊
172	**23**◊
344	**11**◊
688	**5**◊
1,376	2◊
2,752	**1**◊

Find the odd numbers in the halving column (the right column), then get the sum of the partner numbers in the doubling column (the left column). These are highlighted in bold type. The sum of all the bold-faced numbers in the left column gives you the originally required product of 43 and 92. In other words, with the Russian peasant's method we get 43 • 92 = 172 + 344 + 688 + 2,752 = 3,956.

In the previous example, we chose the first column as the doubling column and the second column as the halving column. We could also have done this Russian peasant's method by halving the numbers in the first column and doubling those in the second. See below.

halving	doubling
◇**43**	**92**
◇**21**	**184**
◇10	368
◇**5**	**736**
◇2	1,472
◇**1**	**2,944**

To complete the multiplication, we find the odd numbers in the halving column (in bold type), and then get the sum of their partner numbers in the second column (now the doubling column). This gives us 43 • 92 = 92 + 184 + 736 + 2,944 = 3,956.

You are not expected to do your multiplication in this high-tech era by copying the Russian peasant's method. However it should be fun to observe how this primitive system of arithmetic (an example of a primitive algorithm)* actually does work. Explorations of this kind are not only instructive but entertaining.

Here you see what was done in the above multiplication algorithm.

---

*We have defined an algorithm as a step-by-step problem-solving procedure, especially an established, recursive computational procedure for solving a problem in a finite number of mechanical steps. The way we do division, multiplication, addition, subtraction, and so on mechanically are examples of arithmetic algorithms.

◊ 43 • 92 = (21 • 2 + 1)(92) = 21 • 184 + 92 = 3,956
◊ 21 • 184 = (10 • 2 + 1) (184) = 10 • 368 + 184 = 3,864
10 • 368 = (5 • 2 + 0)(368) = 5 • 736 +   0   = 3,680
◊ 5 • 736 = (2 • 2 + 1)(736) = 2 • 1,472 + 736 = 3,680
2 • 1,472 = (1 • 2 + 0)(1,472) = 1 • 2,944 +   0 = 2,944
◊ 1 • 2,944 = (0 • 2 + 1)(2,944) = 0   + 2,944 = 2,944
                                            3,956

For those familiar with the binary system (i.e., base 2), one can also explain this Russian peasant method with the following representation.

$(43)(92) = (1 \bullet 2^5 + 0 \bullet 2^4 + 1 \bullet 2^3 + 0 \bullet 2^2 + 1 \bullet 2^1 + 1 \bullet 2^0)(92)$
$= 2^0 \bullet 92 + 2^1 \bullet 92 + 2^3 \bullet 92 + 2^5 \bullet 92$
$= 92 + 184 + 736 + 2,944$
$= 3,956$

Whether or not you have a full understanding of the discussion of the Russian peasant's method of multiplication, you should at least now have a deeper appreciation for the common multiplication algorithm you learned in school, even though most people today multiply with a calculator. There are many other multiplication algorithms, yet the one shown here is perhaps one of the strangest and it is through this strangeness that we can appreciate the powerful consistency of mathematics that allows us to conjure up such an algorithm.

2.6. Speed Multiplying by 21, 31, 41, and Others

Much of our arithmetic depends on our use of algorithms, that is, the procedures for doing arithmetic computations. When we learned these algorithms in the early grades of our schooling, we did them without much concern for why they work. In today's technologically advanced world, we find less need for these algorithms. You might enjoy the task of trying to determine why these algorithms actually work. By considering the following "new" algorithm, one you may not have seen before, you have the opportunity to familiarize yourself with an unfamiliar algorithm and then try to see if you can explain why it works. Once you feel comfortable with the algorithm, you can easily extend it to other numbers of this pattern. The examples provided show the algorithm for multiplying by 21, 31, and 41. You should have no trouble extending these to multiplying by 51, 61, and so on. Going through this little exercise should give you better appreciation for the standard algorithms that were required to be learned by rote in your early grades.

To multiply by 21: *Double the number, then multiply by 10 and add the original number.*

For example: to multiply 37 by 21,
Double 37 yields 74, multiply by 10 to get 740, and then add the original number 37 to get 777.

To multiply by 31: *Triple the number, then multiply by 10 and add the original number.*

For example: to multiply 43 by 31,
Triple 43 yields 129, multiply by 10 to get 1290, and then add the original number 43 to get 1,333.

To multiply by 41: *Quadruple the number, then multiply by 10 and add the original number.*

For example: to multiply 47 by 41,
Quadruple 47 yields 188, multiply by 10 to get 1,880, and then add the original number 47 to get 1,927.

By now you should be able to recognize the pattern and might try to extend the rule further.

2.7. Clever addition

One of the most popularly repeated stories from the history of mathematics is the tale of the famous mathematician Carl Friedrich Gauss, who at age ten was said to have mentally added the numbers from 1 to 100 in response to a busy work assignment given by his teacher. Gauss found the answer without writing anything but the answer, while the rest of the class labored without success for another hour.* Although it is a cute story and generally gets a very favorable reaction, it happens to provide for a neat little formula for adding numbers in an arithmetic sequence.

An arithmetic sequence is a list of numbers that has a common difference between consecutive numbers. For example 2, 4, 6, 8, 10, ... is an arithmetic sequence with a common difference of 2. The following are examples of other arithmetic sequences:

5, 10, 15, 20, 25, 30, ... (common difference 5)
7, 11, 15, 19, 23, 27, ... (common difference 4)

Perhaps the simplest arithmetic sequence is: 1, 2, 3, 4, 5, ... These numbers are called the natural numbers.

What Gauss did to get the sum of the first 100 natural numbers without writing a single number was not to add the numbers in the order in which they appear, but rather to add them in the following way:

the first plus the last,
the second plus the next-to-last,
the third plus the third-from-last,
and so on.

If we do this with the sum 1 + 2 + 3 + ... + 98 + 99 + 100, we get the following:

---

*According to E. T. Bell in his book, *Men of Mathematics* (New York: Simon & Schuster, 1937), the problem given to Gauss was of the sort: 81,297 + 81,495 + 81,693 + . . . + 100,899, where the common difference between consecutive terms was 198 and the number of terms was 100. Today's lore uses the numbers to be summed from 1 to 100, which makes the point just as well, but in simpler form.

$$1 + 100 \quad = 101$$
$$2 + 99 \quad = 101$$
$$3 + 98 \quad = 101$$
$$4 + 97 \quad = 101$$
$$\ldots$$
$$50 + 51 \quad = 101$$

Notice that each pair has the same sum of 101. The sum of these fifty pairs of numbers is $50 \bullet 101 = 5{,}050$.

From this example we can derive a useful formula for adding numbers in an arithmetic sequence. In words, we add the first and the last, then multiply this sum by one-half the number of members in the sequence. We can generalize this and get a formula for an arithmetic series of n terms, where a is the first term and l is the last term (using Gauss's method), as follows: Sum $= \frac{n}{2}(a + l)$.

You have here an example of how simple it is to derive a very useful mathematical formula, one based on a very lovely pattern that wasn't completely obvious from the start.

Now let us use the formula to find the sum of the first fifty odd numbers, that is, $1 + 3 + 5 + 7 + \ldots + 99$. Sum $= \frac{50}{2}(1 + 99) = 25(100) = 2{,}500$.

2.8. alphametics

One of the great strides made by Western civilization (which was adopted from the Arabic civilization) was the use of a place value system for our arithmetic. Working with Roman numerals was not only cumbersome but made many algorithms impossible. The first appearance of the Hindu Arabic numerals, as mentioned earlier, was in Fibonacci's book, *Liber abaci* in 1202. Beyond its usefulness, the place value system can also provide us with some recreational mathematics that can stretch our understanding and facility with the place value system and the algorithms we use with it.

Applying reasoning skills to analyzing an addition algorithm situation can be very important in training your mathematical thinking.

This activity may have you strain your number facility. Try it, you'll enjoy it. Consider the following.

The letters below represent the digits of a simple addition if each letter represents a unique digit and M is not equal to zero.

$$
\begin{array}{r}
\text{S E N D} \\
+ \quad \text{M O R E} \\
\hline
\text{M O N E Y}
\end{array}
$$

Find the digits that are represented by the letters to make this addition correct. Show that the solution is unique.

Most important in this activity is the analysis, and particular attention should be given to the reasoning used. At first this may appear to be a daunting task, but if you take it step by step, you will find it rewarding and entertaining.

The sum of two four-digit numbers cannot yield a number greater than 19,998. Therefore **M = 1**.

We then have MORE < 2,000 and SEND < 10,000. It follows that MONEY < 12,000. Thus, O can be either 0 or 1. But the 1 is already used; therefore, **O = 0**.

We now have:

$$
\begin{array}{r}
\text{S E N D} \\
+ \quad \text{1 0 R E} \\
\hline
\text{1 0 N E Y}
\end{array}
$$

Now MORE < 1,100. If SEND were less than 9,000, then MONEY < 10,100, which would imply that N = 0. But this cannot be since 0 was already used; therefore SEND > 9,000, so that **S = 9**.

We now have:

$$
\begin{array}{r}
\text{9 E N D} \\
+ \quad \text{1 0 R E} \\
\hline
\text{1 0 N E Y}
\end{array}
$$

The remaining digits from which we may complete the problem are 2, 3, 4, 5, 6, 7, 8.

Let us examine the units digits. The greatest sum is $7 + 8 = 15$ and the least sum is $2 + 3 = 5$.

If $D + E < 10$, then $D + E = Y$, with no carryover into the tens column. Otherwise $D + E = Y + 10$, with a 1 carried over to the tens column.

Taking this argument one step further to the tens column, we get $N + R = E$, with no carryover, or $N + R = E + 10$, with a carryover of 1 to the hundreds column. However, if there is no carryover to the hundreds column, then $E + 0 = N$, which implies that $E = N$. This is not permissible. Therefore, there must be a carryover to the hundreds column. So $N + R = E + 10$, and $E + 0 + 1 = N$, or $E + 1 = N$.

Substituting this value for N into the previous equation we get: $(E + 1) + R = E + 10$, which implies that $R = 9$. But this has already been used for the value of S. We must try a different approach.

We shall assume, therefore, that $D + E = Y + 10$, since we apparently need a carryover into the tens column, where we just reached a dead end.

Now the tens column's sum is $1 + 2 + 3 < 1 + N + R < 1 + 7 + 8$. If, however, $1 + N + R < 10$, there will be no carryover to the hundreds column, leaving the previous dilemma of $E = N$, which is not allowed. We then have $1 + N + R = E + 10$, which insures the needed carryover to the hundreds column.

Therefore $1 + E + 0 = N$, or $E + 1 = N$.

Substituting this in the above equation $(1 + N + R = E + 10)$ gives us $1 + (E + 1) + R = E + 10$, or **$R = 8$**.

We now have:

$$
\begin{array}{r}
9\,E\,N\,D \\
+\quad 1\ 0\,8\,E \\
\hline
1\,0\,N\,E\,Y
\end{array}
$$

From the remaining list of available digits, we find that $D + E < 14$.

So from the equation $D + E = Y + 10$, Y is either 2 or 3. If $Y = 3$, then $D + E = 13$, implying that the digits D and E can take on only 6 or 7.

If $D = 6$ and $E = 7$, then from the previous equation $E + 1 = N$, we would have $N = 8$, which is unacceptable since $R = 8$.

If D = 7 and E = 8, then from the previous equation E + 1 = N, we would have N = 9, which is unacceptable since S = 9.

Therefore, **Y = 2**.

We now have:

$$
\begin{array}{r}
9\,E\,N\,D \\
+ \quad 1\ 0\,8\,E \\
\hline
1\,0\,N\,E\,2
\end{array}
$$

Thus D + E = 12. The only way to get this sum is with 5 and 7.

If E = 7, we again get, from E + 1 = N, the contradictory N = 8, which is not acceptable.

Therefore, **D = 7** and **E = 5**. We can now again use the equation E + 1 = N to get **N = 6**.

Finally we get the solution:

$$
\begin{array}{r}
9\,5\,6\,7 \\
+ \quad 1\,0\,8\,5 \\
\hline
1\,0\,6\,5\,2
\end{array}
$$

This rather strenuous activity should provide some useful insights into the nature of addition algorithms.

2.9. howlers

In the early years of schooling, we learned to reduce fractions. There were specific ways to do it correctly. Some wise guy seems to have come up with a shorter way to reduce some fractions. Is he right?

He was asked to reduce a fraction and did it in the following way:

$$
\frac{2\cancel{6}}{\cancel{6}5} = \frac{2}{5}
$$

That is, he just canceled out the 6s to get the right answer. Is this procedure correct? Can it be extended to other fractions? If so, then we were surely treated unfairly by our elementary school teachers who made us do much more work. Let's look at what was done here and see if it can be generalized.

In his book, *Fallacies in Mathematics* (Cambridge: Cambridge University Press, 1959), E. A. Maxwell refers to the following cancellations as "howlers":

$$\frac{1\cancel{6}}{\cancel{6}4} = \frac{1}{4} \qquad\qquad \frac{2\cancel{6}}{\cancel{6}5} = \frac{2}{5}$$

Perhaps when someone does the fraction reductions this way and still gets the right answer, it could just make you howl.

Here are two more examples of these howling cancellations:

$$\frac{1\cancel{9}}{\cancel{9}5} = \frac{1}{5}$$

$$\frac{4\cancel{9}}{\cancel{9}8} = \frac{4}{8} = \frac{1}{2}$$

At this point you may be somewhat puzzled. Your first reaction is probably to ask if this can be done to any fraction composed of two-digit numbers of this sort. Can you find another fraction (comprised of two-digit numbers) where this type of cancellation will work? You might cite $\frac{55}{55} = \frac{5}{5} = 1$ as an illustration of this type of cancellation. This will hold true for all two-digit multiples of eleven, but our concern will be only with proper fractions (i.e., whose value is less than one).

For those readers with a good working knowledge of elementary algebra, we can "explain" this situation. That is, why are the four fractions above the *only* ones (composed of two-digit numbers) where this type of cancellation will hold true?

Consider the fraction $\frac{10x+a}{10a+y}$.

The above four cancellations were such that when canceling the a's the fraction was equal to $\frac{x}{y}$.

Therefore, $\frac{10x+a}{10a+y} = \frac{x}{y}$.

This yields:
$$y(10x + a) = x(10a + y)$$
$$10xy + ay = 10ax + xy$$
$$9xy + ay = 10ax$$
$$y(9x + a) = 10ax$$

And so
$$y = \frac{10ax}{9x+a}$$

At this point we shall inspect this equation. It is necessary that x, y, and a are integers since they were digits in the numerator and denominator of a fraction. It is now our task to find the values of a and x for which y will also be integral.

To avoid a lot of algebraic manipulation, you will want to set up a chart which will generate values of y from $y = \frac{10ax}{9x+a}$. Remember that x, y, and a must be single-digit integers. Below is a portion of the table you will be constructing. Notice that the cases where $x = a$ are excluded since $\frac{x}{y} = 1$.

x\a	1	2	3	4	5	6	...	9
1	■	$\frac{20}{11}$	$\frac{30}{12}$	$\frac{40}{13}$	$\frac{50}{14}$	$\frac{60}{15}=4$		$\frac{90}{18}=5$
2	$\frac{20}{19}$	■	$\frac{60}{21}$	$\frac{80}{22}$	$\frac{100}{23}$	$\frac{120}{24}=5$		
3	$\frac{30}{28}$	$\frac{60}{29}$	■	$\frac{120}{31}$	$\frac{150}{32}$	$\frac{180}{33}$		
4				■				$\frac{360}{45}=8$
:								
9								

The portion of the chart pictured above generated the four integral values of y, two of which are: when $x = 1$, $a = 6$ and $y = 4$, and when $x = 2$, $a = 6$ and $y = 5$. These values yield the fractions $\frac{16}{64}$ and $\frac{26}{65}$, respectively. The remaining two integral values of y will be obtained when $x = 1$ and $a = 9$, yielding $y = 5$, and when $x = 4$ and $a = 9$, yielding $y = 8$. These yield the fractions $\frac{19}{95}$ and $\frac{49}{98}$, respectively. This should convince you that there are only four such fractions composed of two-digit numbers which allow this weird digit cancellation to get a properly reduced fraction.

You may now wonder if there are fractions composed of numerators and denominators of more than two digits, where this strange type of cancellation holds true. Try this type of cancellation with $\frac{499}{998}$. You should find that $\frac{499}{998} = \frac{4}{8} = \frac{1}{2}$.

A pattern is now emerging and you may realize that

$$\frac{49}{98} = \frac{499}{998} = \frac{4999}{9998} = \frac{49999}{99998} = \cdots$$

$$\frac{16}{64} = \frac{166}{664} = \frac{1666}{6664} = \frac{16666}{66664} = \frac{166666}{666664} = \cdots$$

$$\frac{19}{95} = \frac{199}{995} = \frac{1999}{9995} = \frac{19999}{99995} = \frac{199999}{999995} = \cdots$$

$$\frac{26}{65} = \frac{266}{665} = \frac{2666}{6665} = \frac{26666}{66665} = \frac{266666}{666665} = \cdots$$

Enthusiastic readers may wish to justify these extensions of the original howlers. Readers who, at this point, have a further desire to seek out additional fractions which permit this strange cancellation should consider the following fractions. They should verify the legitimacy of this strange cancellation and then set out to discover more such fractions.

$$\frac{3\not{3}2}{8\not{3}0} = \frac{32}{80} = \frac{2}{5}$$

$$\frac{3\not{8}5}{8\not{8}0} = \frac{35}{80} = \frac{7}{16}$$

$$\frac{1\not{3}8}{\not{3}45} = \frac{18}{45} = \frac{2}{5}$$

$$\frac{2\not{7}5}{7\not{7}0} = \frac{25}{70} = \frac{5}{14}$$

$$\frac{16\not{3}}{\not{3}26} = \frac{1}{2}$$

Aside from providing an algebraic application, which can be used to introduce a number of important topics in a motivational way, the notion can also provide some recreational activities. Here are some more of these "howlers."

$$\frac{484}{847}=\frac{4}{7}\qquad \frac{545}{654}=\frac{5}{6}\qquad \frac{424}{742}=\frac{4}{7}\qquad \frac{2499}{9996}=\frac{24}{96}=\frac{1}{4}$$

$$\frac{48484}{84847}=\frac{4}{7}\qquad \frac{54545}{65454}=\frac{5}{6}\qquad \frac{42424}{74242}=\frac{4}{7}$$

$$\frac{3243}{4324}=\frac{3}{4}\qquad \frac{6486}{8648}=\frac{6}{8}=\frac{3}{4}$$

$$\frac{14714}{71468}=\frac{14}{68}=\frac{7}{34}\qquad \frac{878048}{987804}=\frac{8}{9}$$

$$\frac{1428577}{4285713}=\frac{1}{3}\qquad \frac{2857742}{8571426}=\frac{2}{6}=\frac{1}{3}\qquad \frac{3461538}{4615384}=\frac{3}{4}$$

$$\frac{7671232877}{8767123287}=\frac{7}{8}\qquad \frac{3243243243}{4324324324}=\frac{3}{4}$$

$$\frac{1025641}{4102564}=\frac{1}{4}\qquad \frac{3243243}{4324324}=\frac{3}{4}\qquad \frac{4571428}{5714285}=\frac{4}{5}$$

$$\frac{4848484}{8484847}=\frac{4}{7}\qquad \frac{5952380}{9523808}=\frac{5}{8}\qquad \frac{4285714}{6428577}=\frac{4}{6}=\frac{2}{3}$$

$$\frac{5454545}{6545454}=\frac{5}{6}\qquad \frac{6923076}{9230768}=\frac{6}{8}=\frac{3}{4}\qquad \frac{4242424}{7424242}=\frac{4}{7}$$

$$\frac{5384615}{7538461}=\frac{5}{7}\qquad \frac{2051282}{8205128}=\frac{2}{8}=\frac{1}{4}\qquad \frac{3116883}{8311688}=\frac{3}{8}$$

$$\frac{6486486}{8648648}=\frac{6}{8}=\frac{3}{4}\qquad \frac{48484848}{84848447}=\frac{4}{7}$$

This unit shows how elementary algebra can be used to investigate a number theory situation, one that is also quite amusing. You see how mathematics continues to hold some hidden treasures.

2.10. that Unusual Number 9

As noted earlier, the first occurrence in western Europe of the Hindu-Arabic numerals we use today was in the book, *Liber abaci*,* written in 1202 by Leonardo of Pisa (otherwise known as Fibonacci). This merchant, enchanted with mathematics, traveled extensively throughout the Arabic countries, where in addition to his business interests, he

---

*Laurence E. Sigler, *Fibonacci's Liber abaci: A Translation into Modern English of Leonardo Pisano's Book of Calculations* (New York: Springer Verlag, 2002).

studied with local mathematicians. In the first chapter of this classic book, he states:

> These are the nine figures of the Indians 9, 8, 7, 6, 5, 4, 3, 2, 1. With these nine figures, and with the symbol, 0, which in Arabic is called zephirum, any number can be written, as will be demonstrated below.

With this book the use of these numerals was first publicized in Europe. Before that, Roman numerals were used. They were, clearly, much more cumbersome to use for calculation.

Fascinated as he was by the arithmetic calculations used in the Islamic world, Fibonacci, in his book, first introduced the system of "casting out nines"* as a check for arithmetic. Even today it still comes in handy. However, the nice thing about it is that it again demonstrates a hidden magic in ordinary arithmetic.

Before we discuss this arithmetic-checking procedure, we will consider how the remainder of a division by 9 compares to removing nines from the digit sum of the number. Let us find the remainder, when 8,768 is divided by 9. The quotient is 974 with a remainder of 2.

This remainder can also be obtained by "casting out nines" from the digit sum of the number 8,768: this means that we will find the sum of the digits and if the sum is more than a single digit we shall repeat the procedure by finding the sum of the digits of this resulting number. In the case of our given number, 8,768, the digit sum is 29 (8 + 7 + 6 + 8 = 29) so we will repeat the process. Again the casting out nines procedure is used to get: 2 + 9 = 11, and again: 1 + 1 = 2, which was the remainder from before, when we divided 8,768 by 9.

Consider the product 734 • 879 = 645,186. We can check this by division,† but that would be somewhat lengthy. We can see if this *could* be correct by "casting out nines." Take each factor in the multi-

---

*"Casting out nines" refers to an arithmetic check that tells you if your answer is possibly correct. The process requires taking bundles of 9s away from the sum of the digits of a number, or subtracting a specific number of 9s from this sum of the digits, or continuously collapsing the number by taking the resulting digits of each digit sum until a single digit remains. All of these will yield the same digit (number).

†By dividing 645,198 by 879 to get 734, if the multiplication is correct.

plicand, multiplier, and product and cast out nines. Add the digits of each, and then continuously add the digits if that sum is not already a single-digit number. When a single-digit number is reached, multiply the two results of the factors to get the digit of the product. Let's apply this procedure now to check the product 645,186 reached above:

For 734: $7 + 3 + 4 = 14$; then $1 + 4 = 5$
For 879: $8 + 7 + 9 = 24$; then $2 + 4 = 6$
For 645,186: $6 + 4 + 5 + 1 + 8 + 6 = 30$

Since $5 \cdot 6 = 30$, which yields 3 (casting out nines: $3 + 0 = 3$), is the same as for the product, the answer *could* be correct.

For practice we will do another casting-out-nines "check" for the following multiplication:

$$56,589 \cdot 983,678 = 55,665,354,342$$

For 56,589: $5 + 6 + 5 + 8 + 9 = 33$; $\longrightarrow 3 + 3 = 6$
For 983,678: $9 + 8 + 3 + 6 + 7 + 8 = 41$; $\longrightarrow 4 + 1 = 5$
For 55,665,354,342: $5 + 5 + 6 + 6 + 5 + 3 + 5 + 4 + 3 + 4 + 2 = 48$; $\blacktriangleright 4 + 8 = 12; 1 + 2 = 3$

To check for possibly having the correct product: $6 \cdot 5 = 30$, $3 + 0 = 3$, which matches the 3 resulting from the product digits.

The same procedure can be used to check the likelihood of a correct sum or quotient, simply by taking the sum (or quotient) and casting out nines, taking the sum (or quotient) of these "remainders" and comparing it with the remainder of the sum (or quotient). They should be equal if the answer is to be correct.

The number 9 has another unusual feature that enables us to use a surprising multiplication algorithm. Although it is somewhat complicated, it is nevertheless fascinating to see it work and perhaps try to determine why this happens. This procedure is intended for multiplying numbers of two digits or more by 9.

It is best to discuss the procedure in context: Multiply 76,354 by 9.

Step 1	Subtract the units digit of the multiplicand from 10	$10 - 4 = 6$
Step 2	Subtract each of the remaining digits (beginning with the tens digit) from 9 and add the result to the previous digit in the multiplicand (for any two-digit sums, carry the tens digit to the next sum.)	$9 - 5 = 4, 4 + 4 = 8$ $9 - 3 = 6, 6 + 5 = 11, 1$ $9 - 6 = 3, 3 + 3 = 6, 6 + 1 = 7$ $9 - 7 = 2, 2 + 6 = 8$
Step 3	Subtract 1 from the left-most digit of the multiplicand	$7 - 1 = 6$
Step 4	List the results in reverse order to get the desired product.	**687,186**

Although it is a bit cumbersome, especially when compared to the calculator, this algorithm provides some insights into number theory. But above all, it's surprising!

2.II. Successive Percentages

Most folks find percentage problems to have long been a nemesis. Problems get particularly unpleasant when multiple percents need to be processed in the same problem. This charmer can turn this one-time nemesis into a delightfully simple arithmetic algorithm that brings with it lots of useful applications and provides new insight into successive percentage problems. This not-very-well-known procedure should enchant you. Let's begin by considering the following problem:

Wanting to buy a coat, Barbara is faced with a dilemma. Two competing stores next to each other carry the same brand coat with the same list price, but with two different discount offers. Store A offers a 10% discount year round on all its goods, but on this particular day offers an additional 20% on top of their already discounted price. Store B simply offers a discount of 30% on that day in order to stay competitive. How many percentage points difference is there between the two options open to Barbara?

At first glance, you may assume there is no difference in price, since $10 + 20 = 30$, yielding the same discount in both cases. Yet with a little more thought you may realize that this is not correct, since in store A only 10% is calculated on the original list price, with the 20% calculated on the lower price, while at store B, the entire 30% is calculated on the original price. Now, the question to be answered is, what percentage difference is there between the discounts in store A and store B?

One expected procedure might be to assume the cost of the coat to be $100, calculate the 10% discount, yielding a $90 price, and an additional 20% of the $90 price (or $18) will bring the price down to $72. In store B, the 30% discount on $100 would bring the price down to $70, giving a discount difference of $2, which in this case is 2%. This procedure, although correct and not too difficult, is a bit cumbersome and does not always allow a full insight into the situation, nor are the numbers always that simple.

An interesting and quite unusual procedure* is provided for entertainment and a fresh look into this problem situation:

Here is a mechanical method (or algorithm) for obtaining a single percentage discount (or increase) in price equivalent to two (or more) successive discounts (or increases) in price.

(1) Change each of the percents involved into decimal form:
.20 and .10

(2) Subtract each of these decimals from 1.00:
.80 and .90 (for an increase in price, add to 1.00)

(3) Multiply these differences:
$(.80)(.90) = .72$

(4) Subtract the above product (i.e., .72) from 1.00:
$1.00 - .72 = .28$, which represents the combined discount
(If the result of step 3 is greater than 1.00, subtract 1.00 from it to obtain the percent of *increase*.)

---

*It is provided without justification of its validity so as not to detract from the solution of the problem. However, for further discussion of this procedure, the reader is referred to A. S. Posamentier and J. Stepelman, *Teaching Secondary School Mathematics: Techniques and Enrichment Units* (Columbus, Ohio: Merrill/Prentice Hall, 6th ed., 2002), pp. 272–74.

When we convert .28 back to percent form, we obtain 28%, the equivalent of successive discounts of 20% and 10%.

This combined percentage of 28% differs from 30% by 2%.

Following the same procedure, you can also combine more than two successive discounts. In addition, successive increases, combined or not combined with a discount, can also be accommodated in this procedure by adding the decimal equivalent of the increase to 1.00, where the discount was subtracted from 1.00, and then continue the procedure in the same way. If the end result comes out greater than 1.00, then this end result reflects an overall increase rather than the discount as found in the above problem.

This procedure not only streamlines a typically cumbersome situation, it also provides some insight into the overall picture. For example, consider the question "Is it advantageous to the buyer in the above problem to receive a 20% discount and then a 10% discount, or the reverse, a 10% discount and then a 20% discount?" The answer to this question is not immediately intuitively obvious. Yet, since the procedure just presented shows that the calculation is merely multiplication, a commutative operation, we find immediately that there is no difference between the two.

So here you have a useful algorithm for combining successive discounts, increases, or combinations of these. Not only is it useful but also it gives you some newfound power in dealing with percentages when using a calculator is not appropriate. What do you think the result will be if a price is increased by 10% and then decreased by 10%? Try using this procedure. The result will surely surprise you!

2.12. are averages averages?

Although baseball batting averages seem to permeate sports discussions, few people realize that these "baseball batting averages" aren't really averages in the true sense of the term: they are actually percentages. Most people, especially after trying to explain this concept, will begin to realize that it is not an average in the way they usually define an "average"—the arithmetic mean. It might be good to search the sports section of the local newspaper to find two baseball players who currently have the same batting average but who have achieved their respective averages with a different number of hits. We shall use a hypothetical example here.

Consider two players: David and Lisa, each with a batting average of .667. David achieved his batting average by getting 20 hits for 30 at bats, while Lisa achieved her batting average by getting 2 hits for 3 at bats.

On the next day, both performed equally, getting 1 hit for 2 at bats (for a .500 batting average); one might expect that they then still have the same batting average at the end of the day. Calculating their respective averages:

David now has 20 + 1 = 21 hits for 30 + 2 = 32 at bats for a $\frac{21}{32} = .656$ batting average.

Lisa now has 2 + 1 = 3 hits for 3 + 2 = 5 at bats for a $\frac{3}{5} = .600$ batting average. Surprise! They do not have equal batting averages.

Suppose we consider the next day, where Lisa performs considerably better than David does. Lisa gets 2 hits for 3 at bats, while David gets 1 hit for 3 at bats. We shall now calculate their respective averages:

David has 21 + 1 = 22 hits for 32 + 3 = 35 at bats for a batting average of $\frac{22}{35} = .629$.

Lisa has 3 + 2 = 5 hits for 5 + 3 = 8 at bats for a batting average of $\frac{5}{8} = .625$.

Amazingly, despite Lisa's superior performance on this day, her batting average, which was the same as David's at the start, is now lower.

There is much to be learned from this "misuse" of the word "average," but, more important, one can get an appreciation of the notion of varying weights of items being averaged.

2.13. the rule of 72

My original premise in this book is to show off mathematics for its own beauty rather than for the power of its applications. This unit essentially does both. I am about to present a method for determining how long it takes to double your money in the bank when it is compounded at a given annual rate. This is clearly good to know, but it is the unusualness of this rule that allows me to exhibit it here. So enjoy it. It is called the "Rule of 72," since it is based on this number. You will soon see.

The "Rule of 72" states that, roughly speaking, *money will double in $\frac{72}{r}$ years when it is invested at an annually compounded interest rate of $r\%$.*

So, for example, if we invest money at an 8% compounded annual interest rate, it will double its value in $\frac{72}{8} = 9$ years. Similarly, if we leave our money in the bank at a compounded rate of 6%, it would take 12 years ($\frac{72}{6} = 12$) for this sum to double its value.

The interested reader might want to better understand why this is so, and how accurate it really is. The following discussion will explain that.

To investigate why or if this really works, we consider the compound interest formula: $A = P(1 + \frac{r}{100})^n$, where A is the resulting amount of money and P is the principal invested for n interest periods at $r\%$ annually.

We need to investigate what happens when $A = 2P$.

The above equation then becomes: $2 = (1 + \frac{r}{100})^n$ (1)

It then follows that $n = \frac{log2}{log(1 + \frac{r}{100})}$.   (2)

Let us make a table of values from the above equation with the help of a scientific calculator:

r	n	nr
1	69.66071689	69.66071689
3	23.44977225	70.34931675
5	14.20669908	71.03349541
7	10.24476835	71.71337846
9	8.043231727	72.38908554
11	6.641884618	73.06073080
13	5.671417169	73.72842319
15	4.959484455	74.39226682

If we take the arithmetic mean (the usual average) of the *nr* values, we get 72.04092314, which is quite close to 72, and so our "Rule of 72" seems to be a very close estimate for doubling money at an annual interest rate of *r*% for *n* interest periods.

An ambitious reader or one with a very strong mathematics background might try to determine a "rule" for tripling and quadrupling money, similar to the way we dealt with the doubling of money. The above equation (2) for *k*-tupling would be $n = \frac{\log k}{\log(1+\frac{r}{100})}$, which for *r* = 8, gives the value for *n* = 29.91884022 (log *k*).

Thus *nr* = 239.3507218 log *k*, which for *k* = 3 (the tripling effect) gives us *nr* = 114.1993167. We could then say that for tripling money we would have a "rule of 114."

However far this topic is explored, the important issue here is that the common "Rule of 72" can be a tool to have at your disposal, and if you have gone far enough in this unit to be able to extend it, great! Remember this useful rule. The rest of this unit is more directed at financial enthusiasts.

2.14. Extracting a Square Root

Why would anyone want to find the square root of a number without using a calculator? Surely, no one would do such a thing. However, it is possible that there are some folks who are curious to know what is actually being done to find a square root. They could develop a better understanding of the process and at least briefly gain some independence from the calculator. We will use a method that was generally not taught in the schools, but gives a good insight into the meaning of a square root. The beauty of this method is that it really allows you to understand what is going on. This method was first published in 1690 by the English mathematician Joseph Raphson (or Ralphson) in his book, *Analysis alquationum universalis*, attributing it to Newton, and therefore the algorithm, the *Newton-Raphson method*, bears both names.

It is perhaps best to see the method used in a specific example: Suppose we wish to find $\sqrt{27}$. Obviously, the calculator would be used here. However, you might like to guess at what this value might be. Certainly it is between $\sqrt{25}$ and $\sqrt{36}$, or between 5 and 6, but closer to 5.

Suppose we guess at 5.2. If this were the correct square root, then if we were to divide 27 by 5.2, we would get 5.2. But this is not the case here, since $\sqrt{27} \neq 5.2$.

We seek a closer approximation. To do that, we find $\frac{27}{5.2} \approx *5.192$. Since $27 \approx 5.2 \bullet 5.192$, one of the factors (5.2 in this case) must be bigger than $\sqrt{27}$ and the other factor (5.192 in this case) must be less than $\sqrt{27}$. Hence, $\sqrt{27}$ is sandwiched between the two numbers 5.2 and 5.192, that is,

$$5.192 < \sqrt{27} < 5.2,$$

so that it is plausible to infer that the average (5.196) is a better approximation for $\sqrt{27}$ than either 5.2 or 5.192.

This process continues, each time with additional decimal places, so that an allowance is made for a closer approximation. That is, since $\frac{5.192+5.196}{2} = 5.194$, then $\frac{27}{5.194} = 5.19831$.

This continuous process provides insight into the finding of the square root of a number, which is not a perfect square.

Cumbersome the method may be, but it surely gives some insight into what a square root represents. Why this neat method wasn't taught in schools in the days before the calculator remains a mystery.

---

*The symbol $\approx$ means "approximately equal to."

3

Problems with Surprising Solutions

When one thinks of mathematics studied in school, it is closely associated with problem solving. Yet the problem solving that comes to mind is related to the tedious and repetitive problems (or exercises) that permeate the typical textbook. This is a very unimaginative form of problem solving. Some people wouldn't even consider that problem solving. Rather, they would call them exercises that require the same technique but with different quantities. True problem solving is a creative and imaginative endeavor that is by no means limited to mathematics. It is done most of our conscious lives. We problem solve when we decide when to get ready to go to the theater, or how best to reach a destination, or how to convince a colleague that our idea is the right one, or how and where to cross a street, or how to solve a mathematical problem. These are all legitimate forms of problem solving. They take on varying degrees of importance, but they all require a strategy for finding a solution.

Problem solving can also be entertaining. That is, if it does not lead to a frustrating experience. What many people find entertaining is when an easily understood problem—one that appears disarmingly very simple—does not allow itself to be easily solved by traditional methods; yet when some unusual approach is used to try to solve it, it becomes surprisingly (almost) trivial. Such is the case with many of the problems offered in this section. They are all easily understood, and when you see the clever solutions, you will be compelled to say "wow!"

Most of the units in this section require nothing more than some very basic high school mathematics. Where necessary, background refreshers are offered. So dive in and let yourself be transformed into a problem-solving lover.

3.1. thoughtful reasoning

Most people, when confronted with a problem, often resort to rather primitive ways of thinking. Sometimes those well trained will consciously think of analogous problems previously solved to see if there is anything that these previous experiences can bring to the current problem. When the primitive methods are used (those that can be called the "peasant's way"), a solution is unlikely. If it emerges, it will have taken considerably more time than an elegant solution (that can be called the "poet's way"), which results from thoughtful reasoning. A nice example of this sort of situation follows. The example shows how unsophisticated thinking does not move you effectively to the solution to a problem.

Given a chessboard and 32 dominos, each the exact size of two of the squares on the chessboard, can you show how 31 of these dominos can cover the chessboard, when a pair of diagonally opposite squares have been removed?

As soon as the question is posed, most people try various arrangements of square covering. This may be done with actual tiles or by drawing a graph grid on paper and then shading adjacent squares two at a time to simulate the tile covering. Before long, frustration begins to set in, since this approach cannot be successful.

Here the issue is to go back to the original question. A careful reading of the question reveals that it does not say to do this tile covering; it asks if it can be done. Yet, because of the way we have been trained, the question is often misread and interpreted as "do it."

A bit of clever insight helps. Ask yourself this question: When a domino tile is placed on the chessboard, what kind of squares are covered? A black square and a white square must be covered by each domino placed on the chessboard. Are there an equal number of black and white squares on the truncated chessboard? No! There are two fewer black squares than white squares. Therefore, it is impossible to cover the truncated chessboard with the 31 domino tiles, since there must be an equal number of black and white squares.

Asking the right questions and inspecting the question asked is an important aspect of being successful in problem solving in mathematics. This unit shows the beauty of mathematical thinking at a very simple, yet profound, level.

3.2. Surprising Solution

Here is a very simple problem with an even simpler solution. Yet the solution most people will come up with is much more complicated. Why? Because they look at the problem in the psychologically traditional way: the way it is presented and played out. Try the problem yourself (don't look below at the solution) and see whether you fall into the "majority-solvers" group. The solution offered later will probably enchant (as well as provide future guidance to) most readers.

A single elimination (one loss and the team is eliminated) basketball tournament has 25 teams competing. How many games must be played until there is a single tournament champion?

Typically the majority-solvers will begin to simulate the tournament (while keeping count of the number of games played), by taking two groups of 12 teams playing the first round, and thereby eliminating 12 teams (12 games have now been played). The remaining 13 teams play, say 6 against another 6, leaving 7 teams in the tournament (18 games have been played now). In the next round, of the 7 remaining teams, 3 can be eliminated (21 games have so far been played). The four remaining teams play, leaving 2 teams for the championship game (23 games have now been played). This championship game is the 24th game.

A much simpler way to solve this problem, one that most people do not naturally come up with as a first attempt, is to focus only on the losers and not on the winners, as we have done above. We ask the key question: How many losers must there be in the tournament with 25 teams in order for there to be one winner? The answer is simple: 24 losers. How many games must be played to get 24 losers? Naturally, 24. So there you have the answer, very simply done. Now, most people will ask themselves, Why didn't I think of that? The answer is, it was contrary to the type of training and experience we have had. Becoming aware of the strategy of looking at the problem from a different point of view may sometimes reap nice benefits, as was the case here. One never knows which strategy will work; just try one and see. In this case, we simply looked at the opposite of what everyone else focuses on—the winners. Looking at the losers gave us a handy way to grasp the solution.

3.3. don't Wine Over this problem

A lengthy reading problem can put some people off for fear that they won't even understand what is being asked. Although this problem does require a fair bit of reading, it is rather easy to understand and could even be dramatized. Once past the statement of the problem, it is very easy to understand, but quite difficult to solve by conventional means.

This is where the beauty of the problem comes in. The elegant solution offered later—as unexpected as it is—almost makes the problem trivial. However, our conventional thinking patterns will likely cause a confusing haze over the problem. Don't despair. Give it a genuine try. Struggle a bit. Then read the solution provided here. We begin by stating the problem.

> We have two one-gallon bottles. One contains a quart of red wine and the other, a quart of white wine. We take a tablespoonful of red wine and pour it into the white wine. Then we take a tablespoon of this new mixture (white wine and red wine) and pour it into the bottle of red wine. Is there more red wine in the white-wine bottle, or more white wine in the red-wine bottle?

To solve the problem, we can figure this out in any of the usual ways—often referred to in the high school context as "mixture problems"—or we can use some clever logical reasoning and look at the problem's solution as follows.

With the first "transport" of wine there is only red wine on the tablespoon. On the second "transport" of wine, there is as much white wine on the spoon as there is red wine in the "white-wine bottle." This may require some readers to think a bit, but most should "get it" soon.

The simplest solution to understand and the one that demonstrates a very powerful strategy is that of *using extremes*. We use this kind of reasoning in everyday life when we resort to the option: "such and such would occur in a worst-case scenario, so we can decide to. . . ."

Let us now employ this strategy for the above problem. To do this, we will consider the tablespoonful quantity to be a bit larger. Clearly the outcome of this problem is independent of the quantity transported. So we will use an *extremely* large quantity. We'll let this quantity actually be the *entire* one quart. That is, following the instructions given in the problem statement, we will take the entire amount (one quart of red wine), and pour it into the white-wine bottle. This mixture is now 50% white wine and 50% red wine. We then pour one quart of this mixture back into the red-wine bottle. The mixture is now the same in both bottles. Therefore, there is as much white wine in the red-wine bottle as there is red wine in the white-wine bottle!

We can consider another form of an extreme case, where the spoon doing the wine transporting has a zero quantity. In this case the conclusion follows immediately: there is as much red wine in the white-wine bottle as there is white wine in the red-wine bottle, that is, zero!

Carefully presented, this solution can be very significant in the way one approaches future mathematics problems and even how one may analyze everyday decision making.

3.4. Working backward

There are many situations where a straightforward approach is by far not the best way to solve a problem. This is true in everyday life situations as well as in mathematics. For example, if you want to see a movie that starts at 7:30 P.M., and you know you have certain things to do before you can be at the movie theater, you would be best off to begin calculating from 7:30 P.M. backward to determine when to start getting ready to leave for the theater. You may figure that it will take you 30 minutes travel time, 1 hour for dinner, 15 minutes to get dressed, and 45 minutes to finish a task you are involved with. This would mean that you should begin to get ready for the theater at 5 P.M.

In mathematics there are lots of examples where working backward is a truly rewarding way to lead you to an elegant solution. One of the best examples of this is a problem that may be a bit "off the beaten path," yet it requires only a little elementary algebra. Let's take a look at the problem.

> If the sum of two numbers is 2 and the product of the same two numbers is 3, find the sum of the reciprocals of the two numbers.

An expected response to this problem is to set up equations that reflect the situation as described above. You might expect to see the following equations:

$$x + y = 2, \text{ and } xy = 3$$

The normal reaction to solving these equations simultaneously is to solve for y in the first equation to get $y = 2 - x$ and substitute this value for y in the second equation. This will then result in the following equation:

$$x(2 - x) = 3, \text{ or}$$
$$x^2 - 2x + 3 = 0$$

This quadratic (as luck would have it) is not factorable. When the quadratic formula* is applied to it, we discover that the roots are imaginary. Then the sum of the reciprocals of these complex numbers is required. The answer will eventually come, but some algebra needs to be dealt with along the way. Although this ought to lead to a correct solution, as you plow along, you will come to the realization that this must be the "peasant's way," and not the "poet's way."

Working backward, the clever alternative, requires that you ask the question: Where will we end up with the solution? Since the sum of the reciprocals is being sought, we must end up with: $\frac{1}{x} + \frac{1}{y}$. Continuing in this spirit, we must ask: What might this have come from? One possibility is the sum of these fractions, namely, $\frac{x+y}{xy}$. A clever person might now realize that we have the solution to the problem staring us right in the face. Remember the two original equations, $x + y = 2$ and

---

*Aquadratic equation is one of the second degree (i.e., one whose highest exponent is 2), and its roots (or solution) can be found by using the quadratic formula: $x = \frac{-b \pm \sqrt{b^2 - 4ac}}{2a}$.

$xy = 3$. They essentially give us the numerator value, 2 and the denominator value, 3 in the fraction $\frac{x+y}{xy}$. So the answer to the original question (problem) is $\frac{2}{3}$. Now isn't that quite a bit simpler than the traditional way? Such dramatic shortcuts are not always available, but the experience of having done this problem in this nifty way will serve you well for future problems that might also be solved more easily using a working-backward strategy.

3.5. logical thinking

When a problem is posed that at first looks a bit daunting, and then a solution—one easily understood—is presented, we often wonder why we didn't think of that simple solution ourselves. It is exactly these problems that should have a dramatic effect on us, and ought to help us with future (analogous) situations. Here is one such problem.

On a shelf in Danny's basement, there are three boxes. One contains only nickels, one contains only dimes, and one contains a mixture of nickels and dimes. The three labels, "nickels," "dimes," and "mixed" fell off, and were all put back on the wrong boxes. Without looking, Danny can select one coin from one of the mislabeled boxes and then must correctly label all three boxes. From which box should Danny select the coin?

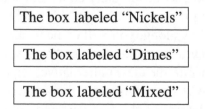

One may reason that the "symmetry" of the problem situation dictates that whatever we can say about the box mislabeled "nickels" could just as well have been said about the box mislabeled "dimes." Thus, if Danny chooses a coin from either of these boxes, the results would be the same.

You should, therefore, concentrate your investigations on what happens if we choose from the box mislabeled "mixed." Suppose Danny selects a nickel from the "mixed" box. Since this box is mislabeled, it cannot be the mixed box and must be, in reality, the nickel box. Since the box marked dimes cannot really be dimes, it must be the "mixed" box. This leaves the third box to be the dimes box. You are probably thinking how simple the problem is—now that you have the solution. It does demonstrate a certain beauty of logical thinking.

3.6. It's Just how You Organize the data

Here is a problem that will draw just a bit on your facility with algebra (very elementary!). When the problem is presented, the symmetry makes it look disarmingly simple, but just wait.

Here is the problem:

Find the numerical value of the following expression:

$$(1 - \tfrac{1}{4})(1 - \tfrac{1}{9})(1 - \tfrac{1}{16})(1 - \tfrac{1}{25})...(1 - \tfrac{1}{225})$$

The usual "knee-jerk" reaction to the question is to get each of the expressions "simplified." That is, doing the indicated subtraction within each pair of parentheses (e.g., $1 - \tfrac{1}{4} = \tfrac{3}{4}$) to get the following:

$$(\tfrac{3}{4})(\tfrac{8}{9})(\tfrac{15}{16})(\tfrac{24}{25})...(\tfrac{224}{225})$$

It is not unusual to then change each fraction to a decimal (with a calculator) and multiply the results (again with a calculator). It is obviously a very cumbersome calculation.

An alternative method would be to organize the data in a different way. This will permit us to look at the problem from a different point of view, with the hope of getting some sort of pattern that will enable us to simplify our work.

$$(1^2 - \tfrac{1}{2^2})(1^2 - \tfrac{1}{3^2})(1^2 - \tfrac{1}{4^2})(1^2 - \tfrac{1}{5^2})...(1^2 - \tfrac{1}{15^2})$$

We now factor each parenthetical expression as the difference of two perfect squares,* which yields:

$$(1 - \tfrac{1}{2})(1 + \tfrac{1}{2})(1 - \tfrac{1}{3})(1 + \tfrac{1}{3})(1 - \tfrac{1}{4})(1 + \tfrac{1}{4})(1 - \tfrac{1}{5})(1 + \tfrac{1}{5})...(1 - \tfrac{1}{15})(1 + \tfrac{1}{15})$$

Now we do the subtraction or addition within each pair of parentheses to get.

$$(\tfrac{1}{2})(\tfrac{3}{2})(\tfrac{2}{3})(\tfrac{4}{3})(\tfrac{3}{4})(\tfrac{5}{4})(\tfrac{4}{5})(\tfrac{6}{5})...(\tfrac{13}{14})(\tfrac{15}{14})(\tfrac{14}{15})(\tfrac{16}{15})$$

A pattern is now evident, and we may multiply pairs of fractions, getting 1 in each case as indicated.

$$(\tfrac{1}{2}) \quad (\tfrac{3}{2})(\tfrac{2}{3}) \quad (\tfrac{4}{3})(\tfrac{3}{4}) \quad (\tfrac{5}{4})(\tfrac{4}{5}) \quad (\tfrac{6}{5})(\tfrac{5}{6})...(\tfrac{14}{13})(\tfrac{13}{14}) \quad (\tfrac{15}{14})(\tfrac{14}{15}) \quad (\tfrac{16}{15})$$

As a result, we are left with $(\tfrac{1}{2})(\tfrac{16}{15}) = \tfrac{8}{15}$.

Here you see the power of a well-organized set of data, which led to a useful pattern.

3.7. focusing on the right information

When faced with a problem with various bits of information, the trick is not to be distracted from the necessary information. Sometimes the trick is not to let the problem situation draw you to the information that will be the less helpful data. Stay focused on what you are asked to find, and let that guide you. This point is perhaps best made with the following problem.

> To extend the amount of wine in a full 16-ounce bottle, Barbara decides upon the following procedure.
> On the first day, she will drink only 1 ounce of the wine and then fill the bottle (i.e., replace the wine) with water.
> On the second day, she will drink 2 ounces of the mixture and then again fill the bottle with water.

---

*The difference of two perfect squares is easily factored as $x^2 - y^2 = (x - y)(x + y)$.

On the third day, she will drink 3 ounces of the mixture and again fill the bottle with water.

She will continue this procedure for succeeding days until she empties the bottle by drinking 16 ounces of the mixture on the 16th day. How many ounces of water will Barbara drink altogether?

It is very easy to get bogged down with a problem like this one. Some attempts will begin by making a table showing the amount of wine and water in the bottle on each day, and attempt to compute the proportional amounts of each type of liquid Barbara drinks on any given day. This will be quite cumbersome.

We could more easily resolve the problem by examining it from another point of view, namely, how much water does Barbara add to the mixture each day? Don't get bogged down with the quantity of wine; this is merely a distractor in this problem situation. Since she eventually empties the bottle (on the 16th day), and it held no water to begin with, she must have consumed all the water that was put into the bottle. So we merely calculate the amount of water Barbara added each time.

On the first day, Barbara added 1 ounce of water.

On the second day, she added 2 ounces of water.

On the third day, she added 3 ounces of water.

On the 15th day, she added 15 ounces of water. (Clearly no water was added on the 16th day.)

Therefore, the number of ounces of water Barbara consumed was what she added each day, namely:

$$1 + 2 + 3 + 4 + 5 + 6 + 7 + 8 + 9 + 10 + 11 + 12 + 13 + 14 + 15 = 120 \text{ ounces.}$$

While this solution is indeed valid, a slightly simpler analogous problem to consider would be to find out how much liquid Barbara drank altogether, and then simply deduct the amount of wine, namely 16 ounces. Thus,

$$1 + 2 + 3 + 4 + 5 + 6 + 7 + 8 + 9 + 10 + 11 + 12 + 13 + 14 + 15 + 16 - 16 = 120 \text{ ounces.}$$

Barbara consumed 136 ounces of liquid, of which 16 ounces was wine and the rest, 120 ounces, must have been water. It is a rather simple problem to solve as long as we focus on the requested information and do not let ourselves get distracted by extraneous information.

3.8. the pigeonhole principle

One of the famous (although often neglected) problem-solving techniques is the so-called pigeonhole principle. In its simplest form, the pigeonhole principle states that if you have, say, 25 items and 24 boxes to put them in, then at least one box must have more than one item in it. In general terms, we in the mathematics community would say that *if you have k + 1 objects that must be put into k holes, then there will be at least one hole with 2 or more objects in it.*

Here is one illustration of the pigeonhole principle at work.

There are 50 teachers' mailboxes in the school's general office. One day the letter carrier delivers 151 pieces of mail for the teachers. After all the letters have been distributed, one mailbox has more letters than any other mailbox. What is the smallest number of letters it can have?

There is a tendency to "fumble around" aimlessly with this sort of problem, usually not knowing where to start. Sometimes, guess and test may work here. However, the advisable approach for a problem of this sort is to consider extremes. Naturally, it is possible for one teacher to get all the delivered mail, but this is not *guaranteed.*

To best assess this situation we shall consider the extreme case, where the mail is as evenly distributed as possible. This would have each teacher receiving 3 pieces of mail with the exception of one teacher, who would have to receive the 151st piece of mail. Therefore, the least number of letters that the box with the most letters received is 4. By the pigeonhole principle, there were 50 3-packs of letters for the 50 boxes. The 151st letter had to be placed into one of those 50 boxes.

3.9. the flight of the bumblebee

Problem solving is not done merely to solve the problem at hand. It is also provided to present various types of problems and, perhaps more important, various procedures for solution. The problem is merely a vehicle for presenting a technique for solution. It is from the types of solutions that one really learns problem solving, since one of the most useful techniques in approaching a problem to be solved is to ask yourself: Have I ever encountered such a problem before? With this in mind, a problem with a very useful "lesson" is presented here. Don't be deterred by the relatively lengthy reading required to get through the problem. You will be delighted (and entertained) with the unexpected simplicity of the solution.

> Two trains serving the Chicago to New York route, a distance of 800 miles, start toward each other at the same time (along the same tracks). One train is traveling uniformly at 60 miles per hour and the other at 40 miles per hour. At the same time, a bumblebee begins to fly from the front of one of the trains, at a speed of 80 miles per hour, toward the oncoming train. After touching the front of this second train, the bumblebee reverses direction and flies toward the first train (still at the same speed of 80 miles per hour). The bumblebee continues this back and forth flying until the two trains collide, crushing the bumblebee. How many miles did the bumblebee fly before its demise?

It is natural to be drawn to find the individual distances that the bumblebee traveled. An immediate reaction is to set up an equation based on the famous (from high school mathematics) relationship: rate times time equals distance. However, this back and forth path is rather difficult to determine, requiring considerable calculation. Just the notion of having to do this can cause serious frustration. Do not allow this frustration to set in. Even if you were able to determine each of the parts of the bumblebee's flight, it is still very difficult to solve the problem in this way.

A much more elegant approach would be to solve a simpler analogous problem (one might also say we are looking at the problem from a different point of view). We seek to find the *distance* the bumblebee traveled. If we knew the *time* the bumblebee traveled, we could determine the bumblebee's distance because we already know the *speed* of the bumblebee. Having two parts of the equation, rate × time = distance, will provide the third part. So having the *time* and the *speed* will yield the distance traveled, albeit in various directions.

The time the bumblebee traveled can be easily calculated, since it traveled the entire time the two trains were traveling toward each other (until they collided). To determine the time the trains traveled, t, we set up an equation as follows: the distance the first train travels is $60t$ and the distance the second train travels is $40t$. The total distance the two trains traveled is 800 miles. Therefore, $60t + 40t = 800$ or $100t = 800$. So $t = 8$ hours, which is also the time the bumblebee traveled. We can now find the distance the bumblebee traveled, using the relationship rate × time = distance, which gives us $80 \times 8 = 640$ miles.

It is important to stress how to avoid falling into the trap of always trying to do what the problem calls for directly. At times a more circuitous method is much more efficient. Lots can be learned from this solution. You see, dramatic solutions are often more useful than traditional solutions, since they provide an opportunity "to think outside the box."

3.10. relating Concentric Circles

More important than the problem itself is the solution method that will be used. More about that later (so as not to spoil the surprise awaiting you).

Consider the following problem:

Two concentric circles are 10 units apart, as shown below. What is the difference between the circumferences of the circles?

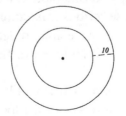

The traditional straightforward method for solving this problem is to find the diameters of the two circles. Then finding the circumference of each circle, we would merely have to subtract to find their difference. Since the lengths of the diameters are not given, the problem is a bit more complicated than usual. Let d represent the diameter of the smaller circle. Then $d + 20$* is the diameter of the larger circle. The circumferences of the two circles will then be πd and $\pi(d + 20)$ respectively.

The difference of the circumferences is

$$\pi(d + 20) - \pi d = \pi d + 20\pi - \pi d = 20\pi.$$

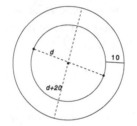

A more elegant and vastly more dramatic procedure would be to use an extreme case. To do this, we will let the smaller of the two circles become smaller and smaller until it reaches an "extreme smallness," and becomes a "point."† In this case, the circle shrunken to a point would become the center of the larger circle. The distance between the two circles now becomes simply the radius of the larger circle. The difference between the lengths of the circumferences of the two circles (required at the start) is now merely the circumference of the larger circle,* or 20π.

Although both procedures yield the same answer, notice how much more work is used for the traditional solution by actually taking the difference of the circumference lengths of the two circles, and how, by considering an extreme situation (without compromising any generality), we reduce the problem to something trivial. Thus, here the beauty of mathematics is manifested in the procedure by which we approach a problem.

---

*That is, the diameter d of the smaller circle plus twice 10, the distance between the circles.

†We can do this because we weren't given the size of either circle. As long as we preserve the length 10, we can allow the two circles to take on any convenient sizes.

3.11. don't Overlook the Obvious

Here is a very entertaining problem that often elicits feelings of self-disappointment (for not having seen such an obvious solution) when the solution is exposed. It is a problem that is certainly solvable when a student has been shown the Pythagorean theorem. As a matter of fact, that knowledge often gets in the way of an elegant solution. Here's the problem:

> The point P is any point on the circle with center O. Perpendicular lines are drawn from P to perpendicular diameters, $\overline{AB}$ and $\overline{CD}$, meeting them at points F and E, respectively. If the diameter of the circle is 8, what is the length of $\overline{EF}$?

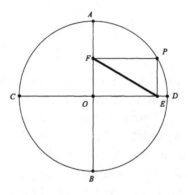

From the training we get, we would most likely look to apply the Pythagorean theorem,* and find that there is no "neat" way to use it. Stepping back from the problem and looking at it in a refreshed way will expose the fact that the quadrilateral $PFOE$ is a rectangle (it was given to have three right angles). Since the diagonals of a rectangle are equal in length, FE must equal PO, which is the radius of the circle and equal to half the diameter, or 4.

   Another way to look at the problem is to take the location of P at a more convenient point, say at point A. In that case, $\overline{FE}$ would coincide with $\overline{AO}$ (since the diameters are perpendicular), which is the radius of the circle.

---

*Since the smaller circle has a circumference of 0.

With either solution you may have been caught off guard. This is not only entertaining but also a good illustration of not overlooking the obvious.

3.12. deceptively difficult (easy)

Here is a problem that looks very simple and is not. It has baffled entire high school mathematics departments! Yet once the solution is shown it becomes quite simple. The result is that you are disappointed in not having seen the solution right from the start. So here it is. Try it without looking at the second diagram. It will give away the solution.

In the figure shown, point E lies on $\overline{AB}$ and point C lies on $\overline{FG}$. The area of parallelogram $ABCD$ = 20 square units. Find the area of parallelogram $EFGD$.†

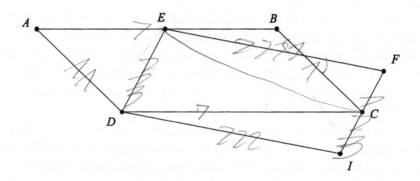

Although the solution is not one that would occur to many people at first thought, the problem can be readily solved using only the tools found in a high school geometry course. Begin by drawing $\overline{EC}$ as in the figure below.

---

*The Pythagorean theorem states that the sides of a right triangle, a, b, and c (where c is the hypotenuse), are in the following relationship: $a^2 + b^2 = c^2$.

†Typically the area of a parallelogram is found by taking the product of its height and base lengths. This is not possible here; so another procedure will have to be found.

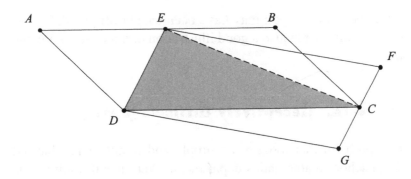

Since triangle *EDC* and parallelogram *ABCD* share a common base ($\overline{DC}$) and a common altitude (a perpendicular from *E* to $\overline{DC}$), the area of triangle *EDC* is equal to one-half the area of parallelogram *ABCD*.

Similarly, since triangle *EDC* and parallelogram *EFGD* share the same base ($\overline{ED}$), and the same altitude to that base (a perpendicular from *C* to $\overline{ED}$), the area of triangle *EDC* equals one-half the area of parallelogram *EFGD*.

Now, since the area of parallelogram *ABCD* and the area of parallelogram *EFGD* are both equal to twice the area of triangle *EDC*, the areas of the two parallelograms must be equal. Thus, the area of parallelogram *EFGD* equals 20 square units, as does parallelogram *ABCD*.

Although the solution method just shown is not often used, it is effective and efficient.

Nevertheless, this problem can be solved quite elegantly by solving a simpler analogous problem (without loss of generality). Recall that the original given conditions were that the two parallelograms had to have a common vertex (*D*), and a vertex of one had to be on the side of the other as shown with points *E* and *C*. Now, let us suppose that *C* coincided with *G*, and *E* coincided with *A*. This satisfies the given condition of the original problem and makes the two parallelograms coincide. Thus, the area of parallelogram *EFGD* = 20 square units.

We could also look at this last solution as one of using extremes. That is, we might consider point *E* on $\overline{AB}$, yet placed at an extreme, such as on point *A*. Similarly, we could place *C* on *G* and satisfy all the conditions of the original problem. Thus, the problem is trivial, in that the two parallelograms coincide.

Remember how difficult you perceived the problem to be at the start?

3.13. the Worst–Case Scenario

Reasoning with extremes is a particularly useful strategy to solve some problems. It can also be seen as a "worst-case scenario" strategy. The best way to appreciate this kind of thinking is through example. So let's experience some really nice reasoning strategies.

> In a drawer, there are 8 blue socks, 6 green socks, and 12 black socks. What is the minimum number of socks Henry must take from the drawer, without looking, to be certain that he has two socks of the same color?

The phrase ". . . *certain* . . . two of the same color," is the key to the problem. The problem does not specify which color, so any of the three would be correct. To solve this problem, you might reason from a "worst-case scenario." Henry picks one blue sock, one green sock, and then one black sock. He now has one of each color, but no matching pair. (True, he might have picked a pair on his first two selections, but the problem calls for "certain.") Notice that as soon as he now picks the fourth sock, it must match the color of one of the previous three socks so that he must have a pair of the same color.

> Consider a second problem:

> In a drawer, there are 8 blue socks, 6 green socks, and 12 black socks. What is the minimum number of socks that Evelyn must take from the drawer (without looking) to be certain that she has two black socks?

Although this problem appears to be similar to the previous one, there is one important difference. In this problem, a specific color has been required. Thus, it is a pair of black socks that we must guarantee being selected. Again, let's use deductive reasoning and construct the "worst-case scenario." Suppose Evelyn first picks all of the blue socks (8). Next she picks all of the green socks (6). Still not one black sock has been chosen. She now has 14 socks in all, but none of them is black. However, the next two socks she picks must be black, since that is the only color remaining. In order to be certain of picking two black socks, Evelyn must select 8 + 6 + 2 = 16 socks in all. Try to create similar problems and present solutions. This can be fun.

4

Algebraic Entertainments

It may not be easy to imagine algebra as a form of entertainment. Some may even feel that algebraic entertainment is an oxymoron! In this chapter, algebra, sometimes called the language of mathematics, is used to make some sense of mathematical phenomena in the behavior of numbers and other areas of mathematics with an eye toward providing entertainment. For example, arithmetic shortcuts are explored, some unusual number relationships are explained, and some beautiful patterns in mathematics are exploited. All of this makes for a rather refreshing use of algebra, which usually manifested itself, in student days, in the form of tedious exercises that often did not appear to be particularly useful. When algebra is used in the school setting, it is used to solve rather routine problems. Here we use algebra to explore other forms of mathematics. For example, the unit on Pythagorean triples gives some very deep insights into these popular groups of numbers.

The goal is to show how algebra can be used to shed new light on and foster a deeper appreciation of mathematical relationships. It ought to entertain you while demonstrating the beauty of algebraic processes.

4.1. Using Algebra to Establish Arithmetic Shortcuts

Suppose you needed to calculate $36^2 - 35^2$. With a calculator this is rather simple to do. But suppose that calculators were not readily available. How would we get the answer in a simple fashion?

We could employ *factoring the difference of two squares*:
$x^2 - y^2 = (x - y)(x + y)$.
That would give us $36^2 - 35^2 = (36 - 35)(36 + 35) = (1)(71) = 71$

The distributive property can be useful for multiplying 8 • 67. Replacing 67 with $(70 - 3)$ allows us to rewrite the multiplication as $8(70 - 3) = 8(70) - 8(3) = 560 - 24 = 536$.

Or to multiply 36 • 14, we can rewrite this as $36(10 + 4) = 36(10) + 36(4) = 360 + 144 = 504$. In the absence of a calculator this is a very efficient way to look at multiplication.

Multiplying two numbers with a difference of 4 can also be simply done by first inspecting the situation in general terms (i.e., algebraically):

The two numbers can be represented as $(x + 2)$ and $(x - 2)$. These have a difference of 4. Their product is $(x + 2)(x - 2) = x^2 - 4$. Thus, we must find the average of the two numbers, x, then square it and subtract 4. For example, to use this idea to multiply 67 • 71, we find the average, 69. Then square 69 to get 4,761, and subtract 4 to get 4,757. It may not always be easier to do the multiplication, but it will give you a sense of "usefulness" of some of the algebra you have been taught.

Multiplying two consecutive numbers uses the property $x(x + 1) = x^2 + x$. This applied to $23 \bullet 24 = 23^2 + 23 = 529 + 23 = 552$ provides a refreshing alternative to the usual multiplication algorithm. Again, it must be stressed that this is not a replacement for the calculator—something most people would not do without.

By this time you might be motivated to discover or establish your own shortcut algorithms. It ought to be fun—though not a calculator replacement!

4.2. the mysterious number 22

At first, this unit will enchant you and then make you wonder why the result is what it is. This is a wonderful opportunity to show off the usefulness of algebra, for it will be through algebra that your curiosity will be quenched.

The reader is urged to work on the following instructions, without checking with the example provided below.

> Select any three-digit number with all digits different from one another. Write all possible two-digit numbers that can be formed from these three digits. Then divide the sum of these two-digit numbers by the sum of the digits in the original three-digit number.

Everyone should get the same answer, 22. This ought to get a big "WOW!"

For example, consider the three-digit number 365. Take the sum of all the possible two-digit numbers that can be formed from these three digits. $36 + 35 + 63 + 65 + 53 + 56 = 308$.

The sum of the digits of the original number is $3 + 6 + 5 = 14$.

Then $\frac{308}{14} = 22$.

For the more ambitious reader, we will analyze this unusual result and we will begin with a general representation of the number:

$100x + 10y + z$.

We now take the sum of all the two-digit numbers constructed from the three digits:

$(10x + y) + (10y + x) + (10x + z) + (10z + x) + (10y + z) + (10z + y)$
$= 22x + 22y + 22z$
$= 22(x + y + z)$

which, when divided by the sum of the digits, $(x + y + z)$, is 22.

    You ought to get a genuine appreciation for algebra with this elementary problem.

    This unit shows the value of algebra in explaining simple arithmetic phenomena and exhibit its beauty.

4.3. justifying an Oddity

In the early years of schooling, we are told that algebra is merely a tool by which we can better understand mathematics. It is then understandable that when we are confronted with a mathematical oddity, it would be nice to have algebra explain the unusual behavior we may be witnessing. This is the case in this unit. Consider this very unusual relationship:

> Any two-digit number ending in 9 can be expressed as the sum of
> the product of the digits and the sum of the digits.

More simply stated: *Any two-digit number ending in 9 = [product of digits] + [sum of digits].*

    Here is what we described above:

$$09 = (0 \bullet 9) + (0 + 9)$$
$$19 = (1 \bullet 9) + (1 + 9)$$
$$29 = (2 \bullet 9) + (2 + 9)$$
$$39 = (3 \bullet 9) + (3 + 9)$$
$$49 = (4 \bullet 9) + (4 + 9)$$
$$59 = (5 \bullet 9) + (5 + 9)$$
$$69 = (6 \bullet 9) + (6 + 9)$$
$$79 = (7 \bullet 9) + (7 + 9)$$
$$89 = (8 \bullet 9) + (8 + 9)$$
$$99 = (9 \bullet 9) + (9 + 9)$$

One of the real advantages of algebra is the facility with which it allows us to justify many mathematical applications.

Don't be too enchanted with the nice pattern; we are using it as a means to an end and not an end in itself. That is, we would like to use this as an illustration of algebra explaining an oddity in mathematics.

We typically represent a two-digit number as $10t + u$, where t represents the tens digit and u represents the units digit. Then the sum of the digits is $t + u$, and the product of the digits is tu.

$$\text{The number meeting the above conditions} = 10t + u = (tu) + (t + u)$$
$$10t = tu + t$$
$$9t = tu$$
$$u = 9 \text{ (for } t \neq 0)*$$

This discussion should evoke a curiosity about numbers with more than two digits. For example:

$$109 = (10 \bullet 9) + (10 + 9)$$
$$119 = (11 \bullet 9) + (11 + 9)$$
$$129 = (12 \bullet 9) + (12 + 9)$$

Here, the digits to the left of the 9 are considered as two-digit numbers and treated just as we treated the tens digit above. The results are the same. This can be extended to any number of digits as long as the units digit is a 9. Have we covered this case with the above algebraic argument? You might want to make the small adjustment in the above algebra to account for this extension.

4.4. Using algebra for number theory

Many unusual number patterns and relationships often boggle the mind. Some cannot be proved (as yet!), such as the famous Goldbach conjecture,† which states:

---

*In this case the rule also holds for $t = 0$.

†Named for Christian Goldbach (1690–1764) and transmitted in a letter to the famous mathematician Leonhard Euler in 1742.

Every even number greater than 2 can be expressed as the sum of two prime numbers.

He also asserted:

Every odd number greater than 5 can be expressed as the sum of three primes.

The latter conjecture, also unproved, is not as widely known as the former. You might want to try verifying these conjectures using a calculator. This will give you a better insight into what is going on here.

Let us now consider a "provable" relationship.

One plus the sum of the squares of any three consecutive odd numbers is always divisible by 12.

The beauty of this is manifested in the simplicity of the procedure used to prove this statement. First, one must establish a way to represent an odd and an even number. For any integer n, $2n$ will always be *even* and $2n + 1$, the next consecutive number, must then be *odd*.

We begin by letting $2n + 1$ be the middle number of the three consecutive odd numbers in consideration. Then, since the preceding odd number is 2 less that this, $[2n + 1] - 2 = 2n - 1$ is the next smaller odd number. Similarly, $[2n + 1] + 2 = 2n + 3$ is the next larger odd number. We are now ready to represent the relationship we are seeking to prove.

$$(2n - 1)^2 + (2n + 1)^2 + (2n + 3)^2 + 1 = 12n^2 + 12n + 12$$
$$= 12(n^2 + n + 1) = 12M, \text{ where } M \text{ represents some integer.*}$$

We can then conclude that this sum of squares plus one is always divisible by 12.

This should be merely a springboard to other similar investigations into number theory.

---

*Since n is an integer, n^2 also is an integer, so the sum $n^2 + n + 1$ also must be an integer. We represent that integer as M, simply for our convenience.

4.5. finding patterns among figurate numbers

We should recall that figurate numbers are those that can be represented by dots in regular polygonal arrangement, as shown below.
There are lots of very surprising relationships that occur with these
numbers. We shall present just a few of these here, with the hope that
you will want to explore further and find some that you can claim as
"your own."

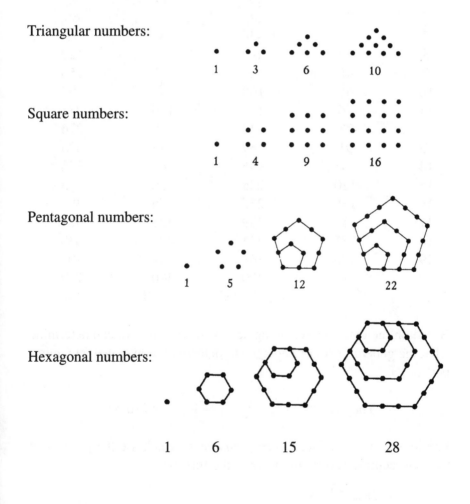

Triangular numbers:

1   3   6   10

Square numbers:

1   4   9   16

Pentagonal numbers:

1   5   12   22

Hexagonal numbers:

1   6   15   28

Consider the following table:

	Triangular Numbers	Square Numbers	Pentagonal Numbers	Hexagonal Numbers
1	1	1	1	1
2	3	4	5	6
3	6	9	12	15
4	10	16	22	28
5	15	25	35	45
6	21	36	51	66
7	28	49	70	91
8	36	64	92	120
9	45	81	117	153
10	55	100	145	190
11	66	121	176	231
12	78	144	210	276
13	91	169	247	325
14	105	196	287	378
15	120	225	330	435
16	136	256	376	496
17	153	289	425	561
18	171	324	477	630
19	190	361	532	703
20	210	400	590	780
n	$\frac{n(n+1)}{2}$ *	$\frac{n(2n-0)}{2} = n^2$	$\frac{n(3n-1)}{2}$	$\frac{n(4n-2)}{2}$

Based on the pattern developing across the nth row, we can determine what the general form of heptagonal, octagonal, nonagonal, etc. numbers are.

They are $\frac{n(5n-3)}{2}$, $\frac{n(6n-4)}{2}$, $\frac{n(7n-5)}{2}$, and so on.

Previously, we introduced *oblong numbers*, which are the product of two consecutive natural numbers in the form:†

---

*This comes from $\frac{n(1n-[-1])}{2} = \frac{n(n+1)}{2}$.

†The product of two consecutive integers.

$$1 \bullet 2 = 2$$
$$2 \bullet 3 = 6$$
$$3 \bullet 4 = 12$$
$$4 \bullet 5 = 20$$
$$5 \bullet 6 = 30, \text{ and so on.}$$

You now have the option of trying to justify the following relationships by using algebra or by convincing yourself through examples that they are, in fact, true. Remember, only a general proof will show they hold for all cases.

An oblong number can be represented as the sum of consecutive even integers, beginning with 2.
Example: $2 + 4 + 6 + 8 = 20$.

An oblong number is twice a triangular number.
Example: $15 \bullet 2 = 30$

*The sum of two consecutive squares and the square of the oblong between them is a square.**
Example: $9 + 16 + 12^2 = 169 = 13^2$

The sum of two consecutive oblong numbers and twice the square between them is a square. †
Example: $12 + 20 + 2 \bullet 16 = 64 = 8^2$

The sum of an oblong number and the next square is a triangular number.
Example: $20 + 25 = 45$

---

*This is a tricky proof, so it is provided for you here. Represent the statement algebraically as $n^2 + (n + 1)^2 + [n(n + 1)]^2$. Expanding and collecting like terms gives us $n^2 + n^2 + 2n + 1 + (n^2 + n)^2 = 2n^2 + 2n + 1 + n^4 + 2n^3 + n^2 = n^4 + 2n^3 + 3n^2 + 2n + 1 = (n^2 + n + 1)^2$. Obviously, we have a square!

†This proof is being provided to show another "trick" in factoring. Consider the algebraic representation: $n(n + 1) + (n + 1)(n + 2) + 2(n + 1)^2$. Now factor out $(n + 1)$:

$(n + 1)[n + (n + 2) + 2(n + 1)] = (n + 1)[4n + 4] = 4(n + 1)^2$, which is a perfect square.

The sum of a square number and the next oblong number is a triangular number.

Example: 25 + 30 = 55

The sum of a number and the square of that number is an oblong number.

Example: 9 + 81 = 90.

Here are some relationships you may wish to try to establish. You might first try to convince yourself that they are true using some specific examples, and then do them algebraically.

- Every odd square number is the sum of eight times a triangular number and 1.
- Every pentagonal number is greater than the sum of three triangular numbers.
- Hexagonal numbers are equal to the odd numbered triangular numbers.

Perhaps you are motivated to find other patterns and then prove that they are true—algebraically, of course.

4.6. Using a Pattern to find the Sum of a Series

Faced with summing a series, most folks would just plow right into the problem, using whatever means they learned to add the terms of the series to find its sum. This could be drudgery. Very inelegant!

Let's look at a situation that lends itself to a few nifty alternatives. Consider the problem of finding the sum of the following series:

$$\frac{1}{1\cdot 2} + \frac{1}{2\cdot 3} + \frac{1}{3\cdot 4} + \ldots + \frac{1}{49\cdot 50}$$

One way to begin is to see if there is any pattern visible. We investigate one possibility:

$$\frac{1}{1\cdot2} = \frac{1}{2}$$

$$\frac{1}{1\cdot2} + \frac{1}{2\cdot3} = \frac{2}{3}$$

$$\frac{1}{1\cdot2} + \frac{1}{2\cdot3} + \frac{1}{3\cdot4} = \frac{3}{4}$$

$$\frac{1}{1\cdot2} + \frac{1}{2\cdot3} + \frac{1}{3\cdot4} + \frac{1}{4\cdot5} = \frac{4}{5}$$

$$\frac{1}{1\cdot2} + \frac{1}{2\cdot3} + \frac{1}{3\cdot4} + \frac{1}{4\cdot5} + \cdots + \frac{1}{n(n+1)} = \frac{n}{n+1}$$

From this pattern, where the sum of each series is an improper fraction formed by the factors of the last fraction in the series,* we will make the guess that the series that goes to $\frac{1}{49\cdot50}$ has the following sum:

$$\frac{1}{1\cdot2} + \frac{1}{2\cdot3} + \frac{1}{3\cdot4} + \frac{1}{4\cdot5} + \cdots + \frac{1}{49\cdot50} = \frac{49}{50},$$

where the numerator and denominator are the factors of the last fraction's denominator.

Another pattern for this series can be obtained by representing each fraction in the series as a difference in the following way:

$$\frac{1}{1\cdot2} = \frac{1}{1} - \frac{1}{2}$$

$$\frac{1}{2\cdot3} = \frac{1}{2} - \frac{1}{3}$$

$$\frac{1}{3\cdot4} = \frac{1}{3} - \frac{1}{4}$$

$$\vdots$$

$$\frac{1}{49\cdot50} = \frac{1}{49} - \frac{1}{50}$$

Adding these equations, the left side gives us our sought-after sum, and on the right side all (except two) of the fractions drop out (with a zero sum, such as $-\frac{1}{2} + \frac{1}{2}$), leaving $\frac{1}{1} - \frac{1}{50} = \frac{49}{50}$.

These very surprising illustrations of useful patterns usually get a reaction such as "Oh, I would never be able to do this on my own," but this should be an unacceptable response, for practice makes perfect!

---

*An *improper fraction* is one where the numerator (the top number) is greater than, or equal to, the denominator (the bottom number).

4.7. geometric View of algebra

The links between algebra and geometry are usually not well known. There are times when the rudiments of algebra can be made concrete by showing that they also make sense from a geometric point of view. More important, it is fun to show the algebraic identities geometrically. After inspecting the ones presented here, you may want to try some for yourself.

The concept of an algebraic identity* can be shown geometrically by using areas to represent second-degree terms and lines to represent first-degree (or linear) terms. Thus a^2 or ab would be represented by an area, and a or b would be represented by a line segment. We will begin by showing the geometric representation of $(a + b)^2 = a^2 + 2ab + b^2$.

To begin, draw a square of side length $(a + b)$. The square should then be partitioned into squares and rectangles, as shown in figure 1. The lengths of the various sides are appropriately labeled.

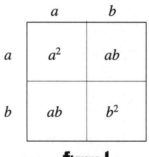

figure I

You should be able to determine the area of each region. Since the area of the large square equals the sum of the areas of the four quadrilaterals into which it was partitioned, you should get

$$(a + b)^2 = a^2 + ab + ab + b^2 = a^2 + 2ab + b^2\dagger$$

---

*An *algebraic identity* is an algebraic equality that is true for all values of the variables.

†Amore rigorous proof can be found in Euclid's *Elements*, Proposition 4, Book II.

Next, illustrate geometrically the identity $a(b + c) = ab + ac$. To begin, draw a rectangle whose adjacent sides are of lengths a and $(b + c)$. The rectangle should then be partitioned into smaller rectangles, as shown in figure 2. The lengths of these sides are labeled.

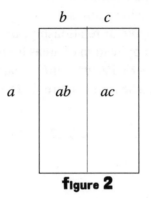

figure 2

It is relatively simple to determine the area of each region. Since the area of the large rectangle equals the sum of the areas of the two quadrilaterals into which it was partitioned, the diagram illustrates $a(b + c) = ab + ac$.

Consider the following identity $(a + b)(c + d) = ac + ad + bc + bd$.

Draw the appropriate rectangle with side lengths $(a + b)$ and $(c + d)$. The rectangle should be partitioned into smaller rectangles (see figure 3). The lengths of the sides and the areas of the smaller regions have been labeled. As in the other cases, the area of the large rectangle equals the sum of the areas of the four quadrilaterals into which it was partitioned.

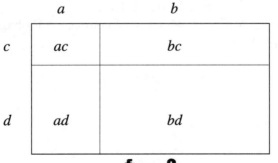

figure 3

The diagram (figure 3) illustrates the identity $(a + b)(c + d) = ac + ad + bc + bd$.

The method of application of areas can be used to prove many algebraic identities. The challenge lies in the choice of dimensions for the quadrilateral and the partitions made.

After you feel comfortable using areas to represent algebraic identities, consider the Pythagorean relationship, $a^2 + b^2 = c^2$. Although this is not an identity, the application of areas is still appropriate. Draw a square of side length $(a + b)$. Partition this square into four congruent triangles and a square, as shown in figure 4. The lengths of the sides have been labeled.

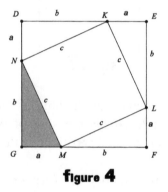

figure 4

The diagram (figure 4) illustrates:

1. Area $DEFG$ = 4(Area of $\triangle GNM$) + Area $KLMN$.
2. Therefore, $(a + b)^2 = 4(\frac{1}{2}ab) + c^2$.
3. If we now substitute the identity for $(a + b)^2$, which was proved above, we obtain, $a^2 + 2ab + b^2 = 2ab + c^2$.

Then clearly $a^2 + b^2 = c^2$, which is, of course, the Pythagorean theorem. Rather neat!

4.8. Some Algebra of the Golden Section

When we talk about the beauty of mathematics, one often thinks of the most beautiful rectangle. This rectangle, often called the Golden Rectangle, has been shown by psychologists to be the most esthetically pleasing rectangle. (See section 5.11.) We will look at this Golden Rectangle from the algebraic point of view—that is, the ratio of its sides, known as the Golden Ratio or Golden Section.*

The Golden Ratio is $\frac{1-x}{x} = \frac{x}{1}$.

By cross-multiplying we get: $x^2 + x - 1 = 0$, and one of the roots is $x = \frac{\sqrt{5}-1}{2}$, for positive x.

We let $\frac{\sqrt{5}-1}{2} = \frac{1}{\phi}$.

Not only does $\phi \cdot \frac{1}{\phi} = 1$ (as would be expected!), but also $\phi - \frac{1}{\phi} = 1$.

This is the only number for which this is true.†

By this time you may want to know what value ϕ has. We can calculate it with the help of a calculator:

$\phi = 1.61803398874989484820458683436563811772030917980576\ldots$
and $\frac{1}{\phi} = 0.61803398874989484820458683436563811772030917980576\ldots$

Notice how the two values differ by 1, which we just stated above.

There are lots of other interesting features of ϕ. For example, consider the following continued fraction:

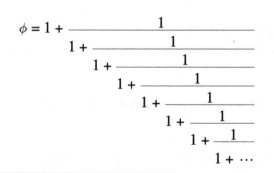

$$\phi = 1 + \cfrac{1}{1 + \cfrac{1}{1 + \cfrac{1}{1 + \cfrac{1}{1 + \cfrac{1}{1 + \cfrac{1}{1 + \cfrac{1}{1 + \cdots}}}}}}}$$

---

*The Gold Ratio refers to the rato of the width of the length of a Golden Rectangle. That is, $\frac{width}{length} = \frac{length}{width+length}$.

†The more ambitious reader can see this verified. Here is the result derived:

Since $\frac{1}{\phi} = \frac{\sqrt{5}-1}{2}$, then $\phi = \frac{2}{\sqrt{5}-1} \cdot \frac{\sqrt{5}+1}{\sqrt{5}+1} = \frac{\sqrt{5}+1}{2}$.

$\phi - \frac{1}{\phi} = \frac{\sqrt{5}+1}{2} - \frac{\sqrt{5}-1}{2} = 1$

To show this, you ought to realize that nothing is lost by truncating the continued fraction at the first numerator. This will give you the following:

$\phi = 1 + \frac{1}{\phi}$, which yields the Golden Ratio.

Another curious relationship is the nest of radicals that follows.

$$\phi = \sqrt{1 + \sqrt{1 + \sqrt{1 + \sqrt{1 + \sqrt{1 + \sqrt{1 + \sqrt{1 + \sqrt{1 + \cdots}}}}}}}}$$

Each of these representations of ϕ is easily verifiable, and can be done with a similar technique. We shall do the second one here and leave the first one for you to do.

$$x = \sqrt{1 + \sqrt{1 + \sqrt{1 + \sqrt{1 + \sqrt{1 + \sqrt{1 + \sqrt{1 + \cdots}}}}}}}$$

Square both sides of the equality.

$$x^2 = 1 + \sqrt{1 + \sqrt{1 + \sqrt{1 + \sqrt{1 + \sqrt{1 + \sqrt{1 + \sqrt{1 + \cdots}}}}}}}$$

Substitute x for the nest of radicals to which it is equal.

$x^2 = 1 + x$

$x^2 - x - 1 = 0$, and then

$x = \phi$, since we defined ϕ at the beginning of this section as a root of this equation.

It is fascinating to observe what happens when we find the powers of ϕ.

$\phi^2 = (\frac{\sqrt{5}+1}{2})^2 = \frac{\sqrt{5}+3}{2} = \frac{\sqrt{5}+1}{2} + 1 = \phi + 1$

$\phi^3 = \phi \bullet \phi^2 = \phi(\phi + 1) = \phi^2 + \phi = (\phi +1) + \phi = 2\phi + 1$

$\phi^4 = \phi^2 \bullet \phi^2 = (\phi + 1)(\phi + 1) = \phi^2 + 2\phi + 1 = (\phi +1) + 2\phi + 1 = 3\phi + 2$

$\phi^5 = \phi^3 \bullet \phi^2 = (2\phi + 1)(\phi + 1) = 2\phi^2 + 3\phi + 1 = 2(\phi +1) + 3\phi + 1 = 5\phi + 3$

$\phi^6 = \phi^3 \bullet \phi^3 = (2\phi + 1)(2\phi + 1) = 4\phi^2 + 4\phi + 1 = 4(\phi +1) + 4\phi + 1 = 8\phi + 5$

$\phi^7 = \phi^4 \bullet \phi^3 = (3\phi + 2)(2\phi + 1) = 6\phi^2 + 7\phi + 2 = 6(\phi +1) + 7\phi + 2 = 13\phi + 8$

and so on.

A summary of this chart reveals a pattern among the coefficients of ϕ (as well among the constants!).

$$\phi^2 = \phi + 1$$
$$\phi^3 = 2\phi + 1$$
$$\phi^4 = 3\phi + 2$$
$$\phi^5 = 5\phi + 3$$
$$\phi^6 = 8\phi + 5$$
$$\phi^7 = 13\phi + 8$$

The coefficients of ϕ and the constants are the Fibonacci numbers. (See section 1.18.)

By this time you are probably thinking that there is no end to the connections that one can draw to the Golden Section. Indeed, you are correct! There are mathematical societies that focus on just these relationships. For us they are a true aspect of beauty in mathematics. To others they represent that and some opportunities for serious mathematics.

4.9. When algebra is not helpful

There are many examples that exhibit the power of algebra. However, there are times when an algebraic solution to a problem is not an advantage. Consider the challenge:

Find four consecutive numbers whose product is 120.

Give yourself a bit of time to begin to tackle this question. Most folks will probably write an algebraic equation to depict the situation. It may look like this:

$$x(x + 1)(x + 2)(x + 3) = 120.$$

Ridding the parentheses here leaves us with a fourth-degree equation (a quartic equation) in one variable.

Rather than to try to solve this quartic equation, a non-algebraic solution might be preferable. Simply guess intelligently and check to get the solution: 2 • 3 • 4 • 5 = 120. You can see from this demonstration that although algebra is very useful to introduce or explain some arithmetic relationships, it is not always the best method. This may come as a shock for some, but that is the beauty of mathematics—to know, through practice, when to use the appropriate techniques available to us.

4.10. rationalizing a denominator

We were taught (in a good algebra class) that it is important for a fraction to have a rational denominator before any arithmetic is done with it, since it is easier to divide by an integer than by an irrational number. Unfortunately, too often rationalizing a denominator was merely an exercise without much purpose. There are applications that show a need for this technique, but somehow these applications are not too convincing about the usefulness of the procedure. There are, however, applications (albeit, somewhat dramatic) that drive home the usefulness argument quite nicely.

Consider the following series for which we are asked to find the sum.

$$\frac{1}{\sqrt{1}+\sqrt{2}} + \frac{1}{\sqrt{2}+\sqrt{3}} + \frac{1}{\sqrt{3}+\sqrt{4}} + \cdots + \frac{1}{\sqrt{2001}+\sqrt{2002}} + \frac{1}{\sqrt{2002}+\sqrt{2003}}$$

You cannot do much with a fraction where the denominator is irrational and so you must seek to change it to an equivalent fraction with a rational denominator. To do this, we multiply each of the fractions by 1 so as not to change its value. The form that 1 should take on is that having the conjugate* of the irrational denominator in both its numerator and denominator.

Rather than "rationalize" the many fractions in the above series, let's do it for a general term to see if that will allow us to convert each of the fractions automatically.

---

*The conjugate of is $a + \sqrt{b}$ is $a - \sqrt{b}$. The product of these is $a^2 - b$, a rational expression.

The general term of this series may be written as $\frac{1}{\sqrt{k}+\sqrt{k+1}}$

We shall now rationalize the denominator of this fraction by multiplying it by 1 in the form of $\frac{\sqrt{k}-\sqrt{k+1}}{\sqrt{k}-\sqrt{k+1}}$ to get

$$\frac{1}{\sqrt{k}+\sqrt{k+1}} \cdot \frac{\sqrt{k}-\sqrt{k+1}}{\sqrt{k}-\sqrt{k+1}} = \frac{\sqrt{k}-\sqrt{k+1}}{-1}.$$

That is, we have found $\frac{1}{\sqrt{k}+\sqrt{k+1}} = \sqrt{k+1} - \sqrt{k}$.

This allows us to rewrite the series as

$$(\sqrt{2}-\sqrt{1})+(\sqrt{3}-\sqrt{2})+(\sqrt{4}-\sqrt{3})+\ldots+(\sqrt{2002}-\sqrt{2001})+(\sqrt{2003}-\sqrt{2002}).$$

By regrouping the terms you will notice that all except two of them drop out by summing to zero, and so we get

$$\sqrt{2003} - 1 \approx 44.754888 - 1 = 43.754888.$$

Here you can see how rationalizing the denominator is not just an exercise without purpose. It is useful, and leads us to a very pretty solution.

4.11. Pythagorean Triples

When the Pythagorean theorem is mentioned, one immediately recalls the famous relationship: $a^2 + b^2 = c^2$. This is probably the most readily recalled topic from the mathematics school experience. Too often there is not enough attention given to the beautiful relationships surrounding this most important theorem. Then while teaching the Pythagorean theorem, teachers often suggest that students recognize (and memorize) certain common ordered triples that can represent the lengths of the sides of a right triangle. Some of these ordered* sets of three numbers, known as *Pythagorean triples*, are

$$(3, 4, 5), \quad (5, 12, 13), \quad (8, 15, 17), \quad (7, 24, 25).$$

*In the parentheses, the order is (a, b, c), where $a^2 + b^2 = c^2$.

How can one find other such Pythagorean triples? This is a topic seldom mentioned in school mathematics—an unfortunate neglect that ought to be rectified. We will do it here to show the algebraic side to the Pythagorean theorem. Warning: The following proofs are a bit challenging, so some readers may wish to skip to the table on page 145.

Suppose you wanted to find out if there is a Pythagorean triple containing 11. Naturally, trial and error may turn up the answer, but this is not very elegant. We shall embark on the adventure of developing a method for establishing some Pythagorean triples. It will give you a nice feeling of doing some real mathematics, while appreciating the beauty of the style of approach.

Before beginning to develop formulas, we must consider a few simple "lemmas" (these are "helper" theorems).

Lemma 1: When the square of an odd number is divided by 8, the remainder is 1.

Proof: We can represent an odd number by $2k + 1$, where k is an integer. The square of this number is $(2k + 1)^2 = 4k^2 + 4k + 1 = 4k(k + 1) + 1$.

Since k and $k + 1$ are consecutive, one of them must be even. Therefore $4k(k + 1)$ must be divisible by 8. Thus $(2k + 1)^2$, when divided by 8, leaves a remainder of 1.

The next lemmas follow directly.

Lemma 2: When the sum of two odd square numbers is divided by 8, the remainder is 2.

Lemma 3: The sum of two odd square numbers cannot be a square number.

Proof: Since the sum of two odd square numbers, when divided by 8, leaves a remainder of 2, the sum is even, but not divisible by 4. It therefore cannot be a square number.

We are now ready to begin our development of formulas for Pythagorean triples. Let us assume that (a, b, c) is a primitive*

---

*A *primitive* Pythagorean triple is one with no common factor aside from 1.

Pythagorean triple. This implies that a and b are relatively prime.*
Therefore they cannot both be even. Can they both be odd?

If a and b are both odd, then by lemma 3: $a^2 + b^2 \neq c^2$. This contradicts our assumption that (a, b, c) is a Pythagorean triple; therefore a and b cannot both be odd. Therefore one must be odd and one even.

Let us suppose that a is odd and b is even. This implies that c must be odd. We can rewrite $a^2 + b^2 = c^2$ as

$$b^2 = c^2 - a^2$$
$$b^2 = (c + a)(c - a)$$

Since the sum and difference of two odd numbers is even, $c + a = 2p$ and $c - a = 2q$ (where p and q are natural numbers).†

By solving for a and c we get

$$c = p + q \text{ and } a = p - q$$

We can now show that p and q must be relatively prime. Suppose p and q were not relatively prime; say g (where $g > 1$) is a common factor. Then g would also be a common factor of a and c. Similarly, g would also be a common factor of $c + a$ and $c - a$. This would make g^2 a factor of b^2, since $b^2 = (c + a)(c - a)$. It follows that g would then have to be a factor of b. Now if g is a factor of b and also a common factor of a and c, then a, b, and c are not relatively prime. This contradicts our assumption that (a, b, c) is a *primitive* Pythagorean triple. Thus, p and q must be relatively prime.

Since b is even, we may represent b as $b = 2r$
But $b^2 = (c + a)(c - a)$.
Therefore $b^2 = (2p)(2q) = 4r^2$, where it follows that $pq = r^2$

If the product of two relatively prime natural numbers (p and q) is the square of a natural number (r), then each of them must be the square of a natural number.

---

*Relatively prime means that they do not have any common factor aside from 1. If a and b had a common factor, then $a^2 + b^2$ would also have a common factor, which would then mean a, b, c had a common factor—a contradiction.

†Natural numbers are our counting numbers: 1, 2, 3, 4, 5 . . .

Therefore we let $p = m^2$ and $q = n^2$, where m and n are natural numbers. Since they are factors of relatively prime numbers (p and q), they (m and n) are also relatively prime.

Since $a = p - q$ and $c = p + q$, it follows that $a = m^2 - n^2$ and $c = m^2 + n^2$.

Also, since $b = 2r$ and $b^2 = 4r^2 = 4pq = 4m^2n^2$, $b = 2$ mn.

To summarize, we now have formulas for generating Pythagorean triples:

$$a = m^2 - n^2 \qquad b = 2mn \qquad c = m^2 + n^2$$

The numbers m and n cannot both be even, since they are relatively prime.* They cannot both be odd, for this would make $c = m^2 + n^2$ an even number, which we established earlier as impossible. Since this indicates that one must be even and the other odd, $b = 2mn$ must be divisible by 4. Therefore, no Pythagorean triple can be composed of three prime numbers. This does *not* mean that the other members of the Pythagorean triple may not be prime.

Let us reverse the process for a moment. Consider relatively prime numbers m and n ($m > n$), where one is even and the other odd. We will now show that (a, b, c) is a primitive Pythagorean triple, where $a = m^2 - n^2$, $b = 2mn$, and $c = m^2 + n^2$.

It is simple to verify algebraically that

$$(m^2 - n^2)^2 + (2mn)^2 = (m^2 + n^2)^2,$$

thereby making it a Pythagorean triple. What remains is to prove that (a, b, c) is a *primitive* Pythagorean triple.

Suppose a and b have a common factor $h > 1$. Since a is odd, h must also be odd. Because $a^2 + b^2 = c^2$, h would also be a factor of c. We also have h a factor of $m^2 - n^2$ and $m^2 + n^2$ as well as of their sum, $2m^2$, and their difference, $2n^2$.

Since h is odd, it is a common factor of m^2 and n^2. However, m and n (and as a result, m^2 and n^2) are relatively prime. Therefore, h cannot be a common factor of m and n. This contradiction establishes that a and b are relatively prime.

---

*This enables us to get *primitive* Pythagorean triples, which are Pythagorean triples that do not have a common factor other than 1.

Having finally established a method for generating primitive Pythagorean triples, you should be eager to put it to use. The following table gives some of the smaller primitive Pythagorean triples.

Some Primitive Pythagorean Triples

m	n	a	b	c
2	1	3	4	5
3	2	5	12	13
4	1	15	8	17
4	3	7	24	25
5	2	21	20	29
5	4	9	40	41
6	1	35	12	37
6	5	11	60	61
7	2	45	28	53
7	4	33	56	65
7	6	13	84	85

A fast inspection of the table indicates that certain primitive Pythagorean triples (a, b, c) have $c = b + 1$. Can you discover the relationship between m and n for these triples?

You may notice that for these triples $m = n + 1$. To prove this will be true for other primitive Pythagorean triples (beyond those in the table), let $m = n + 1$ and generate the Pythagorean triples.

$$a = m^2 - n^2 = (n + 1)^2 - n^2 = 2n + 1$$
$$b = 2mn = 2n(n + 1) = 2n^2 + 2n$$
$$c = m^2 + n^2 = (n + 1)^2 + n^2 = 2n^2 + 2n + 1$$

Clearly $c = b + 1$, which was to be shown!

A natural question may be to find all primitive Pythagorean triples that are consecutive natural numbers. In a method similar to that used above, you ought to find that the only triple satisfying that condition is $(3, 4, 5)$.

From this rather lengthy development you should have a far better appreciation for Pythagorean triples and elementary number theory. Here are some other investigations that the more ambitious reader may wish to try.

1. Find six primitive Pythagorean triples that are not included in the table.
2. Prove that every primitive Pythagorean triple has one member that is divisible by 3.
3. Prove that every primitive Pythagorean triple has one member that is divisible by 5.
4. Prove that for every primitive Pythagorean triple the product of its members is a multiple of 60.
5. Find a Pythagorean triple (a, b, c) where $b^2 = a + 2$.

5

Geometric Wonders

this chapter is larger than the others since the visual effect of geometry lends itself particularly well to exhibiting the beauty of mathematics. Most of the units require just a very basic recollection of high school geometry (without proofs, of course!). Those that appear to require some level of geometric sophistication can also be treated in a more elementary fashion.

There are a number of units that demonstrate the beautiful concept of an *invariant* in geometry. This means that, in some situations, critical aspects of a figure remain constant even when some parts are changed. That is, suppose we can accept the fact that the altitudes of a triangle are concurrent (intersect in one point); we can then draw any shape triangle and this will still be true. The relationship is invariant regardless of the kind of triangle used. These invariants can be nicely demonstrated on the computer with the help of the Geometer's Sketchpad* program. For example, the perpendiculars drawn to the three sides (or their extensions) of a triangle from any point on its circumcircle† intersect the sides in three collinear points (Simson's invariant).‡ That is beautifully demonstrated with the Geometer's Sketchpad. Such an invariant is just one of several shown in this chapter.

There are several very entertaining proofs of the Pythagorean theorem, one by paper folding, one extraordinarily simple, and one done by former U.S. President James A. Garfield. There are units that will require the reader to do some activities, such as "moving" a pencil along a figure, and measuring a tall drinking glass, while exposing some extraordinary properties/phenomena.

The chapter is full of unusual geometric properties, all pointing to the beauty of the subject matter. It is for the reader to reap the most out of this chapter. In many cases further investigation beyond these pages may be illuminating.

5.1. angle Sum of a triangle

It is widely accepted that for a plane triangle the sum of the measures of the angles is 180°. This by no means insures that everyone knows what that really means. This basis for Euclidean geometry ought to be genuinely understood by all. Most people know that when they make one complete revolution, that that represents 360°. There is nothing sacrosanct about this measure, other than that it is generally accepted and so used. That is, we make an agreement that a complete revolution is 360°. So a half revolution is 180°.

How does the angle sum of a triangle relate to this? The simplest and perhaps the most convincing way to demonstrate this angle sum is to tear the three vertices from a paper triangle and place them together to form the straight line. The straight line represents one-half of a complete revolution, hence, the 180°.

---

*Geometer's Sketchpad is a very versatile computer software program published by Key Curriculum Press, www.keypress.com.

†The circumcircle of a polygon is a circle containing all the polygon's vertices.

‡Collinear points are points that lie on the same straight line.

It is perhaps neater to use a folding procedure. Cut a conveniently large scalene triangle* from a piece of paper. Then fold one vertex so that it touches the opposite side and so that the crease is parallel to that side (see figure 1).

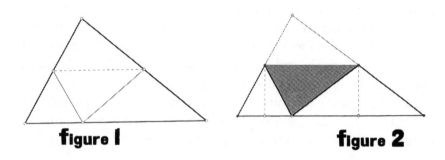

figure 1   **figure 2**

Then fold the remaining two vertices to meet the first vertex at a common point (figure 2). You will notice that the three angles of the triangle come together and form a straight line, and hence have an angle sum of 180° (figure 3).

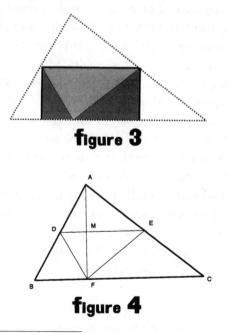

figure 3

figure 4

---

*Ascalene triangle is one where all sides have different lengths.

However, it is also nice to show why this folding procedure has the vertices meet at a point on the side of the triangle. Establishing this phenomenon is tantamount to proving the theorem establishing the angle sum of a triangle.

The proof of this theorem for the interested reader follows directly from the paper-folding exercise using figure 4. By folding the top vertex along a parallel crease* (i.e., $\overline{DE}\|\overline{BC}$), $\overline{AF}\perp\overline{ED}$ at M. Since $\overline{MF}\cong\overline{AM}$, or M is the midpoint of $\overline{AF}$, D and E are midpoints of $\overline{AB}$ and $\overline{AC}$, respectively, since a line parallel to one side of a triangle (either $\triangle BAF$ or $\triangle CAF$) and bisecting a second side ($\overline{AF}$) of the triangle, also bisects the third side. It is then easy to show that since $\overline{AD}\cong\overline{DF}$, $\overline{DB}\cong\overline{DF}$, and similarly $\overline{EF}\cong\overline{EC}$, so that the folding over of vertices B and C would fit at F, forming a straight line along $\overleftrightarrow{BFC}$.

5.2. pentagram angles

The pentagram is one of the most interesting figures in geometry. The Pythagoreans even used it as their symbol. A regular pentagram contains the Golden Ratio (which you will learn more about later), and in that shape it adorns the American flag fifty times!

We know that the sum of the angles of a triangle is 180° and a quadrilateral's angles have a sum of 360°. But what is the sum of the angles of a pentagram? Although easily provable, we shall assume that all pentagrams have the same angle sum. This implies that we ought to be able to get the answer by finding the angle sum of a regular pentagram and then simply generalize it to all pentagrams. You ought to be able to "stumble" on this angle sum once you have been able to find the measure of one vertex angle, not very difficult since the angles are all congruent and there is lovely symmetry throughout.

However, suppose we didn't make this connection, and were simply trying to get the angle sum of an "ugly," arbitrarily drawn pentagram, such as the following diagram.

---

*To refresh your memory, the following symbols are defined: ‖ means "is parallel to," ⊥ means "is perpendicular to," ≅ means "is congruent to," and ~ means "is similar to."

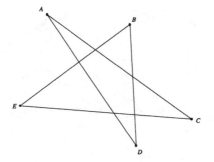

We could determine this by placing a pencil on $\overline{AC}$ in the direction pointing at A, and rotating it through $\angle A$, so that it is now on $\overline{AD}$ pointing at A. Then rotate it through $\angle D$ so that it is now on $\overline{BD}$ pointing at B. Then rotate it through $\angle B$ so that it is now on $\overline{BE}$ pointing at B. Then rotate it through $\angle E$ so that it is now on $\overline{EC}$ pointing at C. Last, rotate it through $\angle C$ so that it is now on $\overline{AC}$ pointing at C, which is the opposite direction of its starting position. Therefore, the pencil reversed its direction, which is the same thing as a rotation of 180°, implying that the angle sum of the pentagram (through which the pencil was rotated, angle by angle) is 180°.

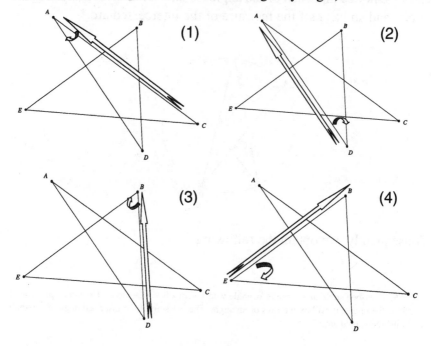

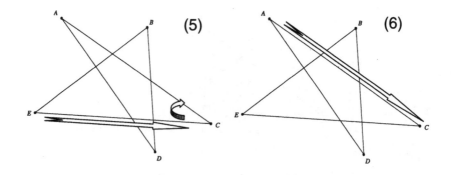

Again, notice how, through the sequence of angle moves, the pencil's direction changed by 180°.

For those who feel more comfortable with a geometric "proof," the following demonstration is provided. Note that we are accepting the notion that the angle sum of the corners of a pentagram is the same for all pentagrams. Since the type of pentagram was not specified, we can assume the pentagram to be regular, or that it is one which is inscribable in a circle (i.e., all of its vertices lie on the circle). In either case, we notice that each of the angles is now an inscribed angle of the circle, and so has half the measure of the intercepted arc.*

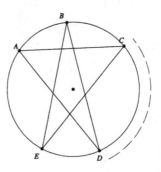

Consequently, we obtain the following:

---

*An inscribed angle of a circle is an angle whose vertex is *on* the circle. The *intercepted arc* of a circle is the one "cut off" by the rays of an angle. The measure of the inscribed angle is one-half that of its intercepted arc.

$m\angle A = \frac{1}{2}m\ \overset{\frown}{CD}$ [this reads "measure of angle A equals one-half the measure of arc CD"]
$m\angle B = \frac{1}{2}m\ \overset{\frown}{ED}$
$m\angle C = \frac{1}{2}m\ \overset{\frown}{AE}$
$m\angle D = \frac{1}{2}m\ \overset{\frown}{AB}$
$m\angle E = \frac{1}{2}m\ \overset{\frown}{BC}$

If we add these equalities, we obtain
$$m\angle A + m\angle B + m\angle C + m\angle D + m\angle E =$$
$$\tfrac{1}{2}m(\overset{\frown}{CD} + \overset{\frown}{ED} + \overset{\frown}{AE} + \overset{\frown}{AB} + \overset{\frown}{BC}) = \tfrac{1}{2} \bullet 360° = 180°$$

That is, the sum of the measures of the angles of the pentagram is one-half the degree measure of the circle, or 180°. Again, note that there was no loss of generality by allowing the nonspecified pentagram to assume a more useful configuration. In other words, we were able to change the figure to one that is more manageable and still lose nothing in the process.

5.3. Some mind-bogglers on π

From early exposure to mathematics, we become familiar with π. Since the most popular formulas in elementary mathematics (and those that seem to stick with us long after we really know what they mean) are $2\pi r$ and πr^2, we may begin to lose sight of what π means and may need reminding. The best way to accomplish this is to show something a bit dramatic. Perhaps the following "experiment" will do the trick.

Take a tall and narrow cylindrical drinking glass. Ask a friend if the circumference is greater or less than the height. The glass should be chosen so that it would "appear" to have a longer height than its circumference. (The typical tall narrow drinking glass fits this requirement.) Now ask your friend how she might test her conjecture (aside from using a piece of string). Recall for her that the formula for the circumference of a circle is $C = \pi d$ (π times the diameter). She may recall that $\pi = 3.14$ is the usual approximation, but we'll be even more crude and use $\pi = 3$. Thus the circumference will be 3 times the diameter, which can be easily "measured" with a stick or a pencil and then marked off 3 times along the height of the glass. Usually you will find that the circumference is longer than the height of the tall glass, even though it does not "appear" to be so. This little optical trick is useful to demonstrate the value of π.

Now for a real mind blower! To appreciate the next revelation on π, you need to know that until recently virtually all the books on the history of mathematics stated that in its earliest manifestation in history, namely the Bible (Old Testament), its value is given as 3. Yet recent detective work shows otherwise.*

One always relishes the notion that a hidden code can reveal long-lost secrets. Such is the case with the common interpretation of the value of π in the Bible. There are two places in the Bible where the same sentence appears, identical in every way except for one word, spelled differently in the two citations. The description of a pool, or fountain, in King Solomon's temple is referred to in the passages that may be found in 1 Kings 7:23 and 2 Chron. 4:2, and read as follows:

> And he made the molten sea of ten cubits from brim to brim, round
> in compass, and the height thereof was five cubits; and *a line* of
> thirty cubits did compass it round about.

The circular structure described here is said to have a circumference of 30 cubits and a diameter of 10 cubits. (A cubit is the length from a person's fingertip to his elbow.) From this we notice that the Bible has $\pi = \frac{30}{10} = 3$. This is obviously a very primitive approximation of π. A late-eighteenth-century rabbi, Elijah of Vilna (Poland), one of the great modern biblical scholars who earned the title "Gaon of Vilna" (meaning brilliance of Vilna), came up with a remarkable discovery, one that could make most history-of-mathematics books faulty if they say that the Bible approximated the value of π as 3. Elijah of Vilna noticed that the Hebrew word of "line measure" was written differently in each of the two biblical passages mentioned above.

In 1 Kings 7:23 it was written as קוה, whereas in 2 Chron. 4:2 it

*Alfred S. Posamentier and Noam Gordon, "An Astounding Revelation on the History of π," *Mathematics Teacher* 77 (January 1984): 52.

was written as קו. Elijah applied the biblical analysis technique (still used today) called gematria, where the Hebrew letters are given their appropriate numerical values according to their sequence in the Hebrew alphabet, to the two spellings of the word for "line measure" and found the following.

The letter values are ק = 100, ו = 6, and ה = 5. Therefore, the spelling for "line measure" in 1 Kings 7:23 is קוה = 5 + 6 + 100 = 111, while in 2 Chron. 4:2 the spelling קו = 6 + 100 = 106. He then took the ratio of these two values: $\frac{111}{106} = 1.0472$ (to four decimal places), which he considered the necessary correction factor, for when it is multiplied by 3, which is believed to be the value of π stated in the Bible, one gets 3.1416, which is π correct to four decimal places! "Wow!!!" is a typical reaction. Such accuracy is quite astonishing for ancient times. To support this notion, take a string to measure the circumference and diameter of several circular objects and find their quotient. You will most likely not get near this four-place accuracy. Moreover, to really push the point of the high degree of accuracy of four decimal places, chances are if you took the average of all your π measurements, you still probably wouldn't get to four-place accuracy.

5.4. the ever-present parallelogram

To properly convince yourself of the power of a geometric theorem (or property), you might begin by drawing a few "ugly" (i.e., any shaped) quadrilaterals. Then (very carefully) locate the midpoints of the four sides of each quadrilateral. If you join these points consecutively, each drawing should result in a parallelogram. Wow! How did this happen? You began (most likely) with different shaped quadrilaterals. Yet every quadrilateral ended up with a parallelogram inside it.

Here are a few possible results:

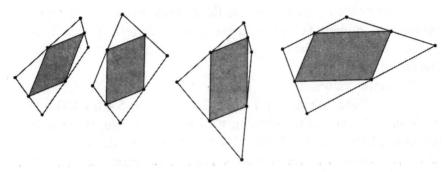

A question that ought to be asked at this point is how might the original quadrilateral have been shaped for the parallelogram to be a rectangle, rhombus, or square?

Either through trial and error or by an analysis of the situation, you should discover the following:

When the diagonals of the original quadrilateral are perpendicular, the parallelogram is a *rectangle*.

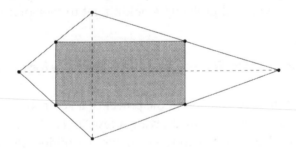

When the diagonals of the original quadrilateral are congruent, then the parallelogram is a *rhombus*.

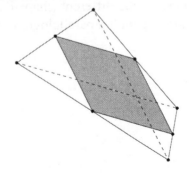

When the diagonals of the original quadrilateral are congruent and perpendicular, then the parallelogram is a *square*.

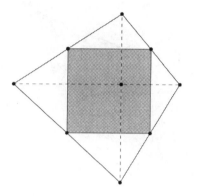

For the more ambitious reader, Geometer's Sketchpad software is highly recommended. For those who wish to prove that all of the above is "really true," a short proof outline is provided, one that should be within easy reach for a high school geometry student.

Proof outline:
The proof is based on a simple theorem that states that a line segment joining the midpoints of two sides of a triangle is parallel to and half the length of the third side of the triangle. This is precisely what happens here.

In $\triangle ADB$, the midpoints of sides $\overline{AD}$ and $\overline{AB}$ are F and G, respectively.
Therefore, $\overline{FG} \| \overline{DB}$ and $FG = \frac{1}{2}BD$, and $\overline{EH} \| \overline{DB}$ and $EH = \frac{1}{2}BD$.
Therefore, $\overline{FG} \| \overline{EH}$ and $FG = EH$. This establishes $FGHE$ as a parallelogram.

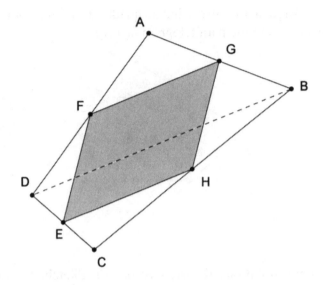

Furthermore, if the diagonals, $\overline{DB}$ and $\overline{AC}$, are congruent, then the sides of the parallelogram must also be congruent, since they are each one-half the length of the diagonals of the original quadrilateral. This results in a rhombus.

Similarly, if the diagonals of the original quadrilateral are perpendicular and congruent, then since the sides of the parallelogram are, in pairs, parallel to the diagonals and half their length, the adjacent sides of the parallelogram must be perpendicular and congruent to each other, making it a square.

5.5. Comparing Areas and Perimeters

Comparing areas and perimeters is a very tricky thing. A given perimeter can yield many different areas. And a given area can be encompassed by many different perimeters. For example, rectangles of perimeter 20 may have very different areas.

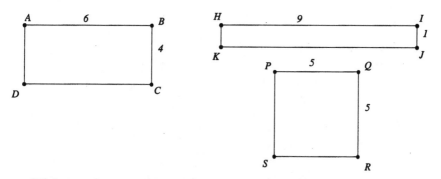

With a perimeter of 20, the area of rectangle *ABCD* is 24.

With a perimeter of 20, the area of rectangle *HIJK* is 9.

With a perimeter of 20, the area of rectangle *PQRS* is 25.

It can be shown that the *maximum* area of a rectangle with a fixed perimeter is the one with equal length and width, that is, a square.

Just the opposite is true for a rectangle of given area, in which the *minimum* perimeter is that where the length and width are equal, that is, a square.

It is interesting to compare areas of similar figures. We will consider circles.

Suppose you have four equal lengths of string.

With the *first piece of string*, one circle is formed.

The *second piece of string* is cut into two equal parts and two congruent circles are formed.

The *third piece of string* is cut into three equal pieces and three congruent circles are formed.

In a similar way, four congruent circles are formed from the *fourth piece of string*.

They are shown below. Note that the sum of the circumferences of each group of congruent circles is the same.

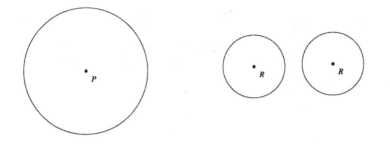

Circle	Diameter	Each circle's circumference	Sum of the circles' circumferences	Each circle's area	Sum of the circles' areas
P	12	12π	12π	36π	36π
R	6	6π	12π	9π	18π
Q	4	4π	12π	4π	12π
S	3	3π	12π	2.25π	9π

An inspection of the above chart shows that the sum of the circumferences for each group of circles is the same, yet the sum of the areas is quite different. The more circles we formed with the same total length of string the smaller the total area of the circles. Just what you would *not* expect to happen!

That is, when two equal circles were formed, the total area of the two circles was one-half that of the large circle. Similarly, when four equal circles were formed, the total area of the four circles was one-fourth of the area of the large circle.

This seems to go against one's intuition. Yet if we consider a more extreme case, with 100 smaller equal circles, we would see that the area of each circle becomes extremely small and the *sum* of the areas of these 100 circles is one-hundredth of the area of the larger circle.

Try to explain this rather disconcerting concept. It ought to give you an interesting perspective on comparison of areas.

5.6. how eratosthenes measured the Circumference of the earth

Measuring the earth today is not terribly difficult, but thousands of years ago this was no mean feat. Remember, the word "geometry" is derived from "earth measurement." Therefore it is appropriate to consider this issue in one of its earliest forms. One of these measurements of the circumference of the earth was made by the Greek mathematician Eratosthenes, in about 230 B.C.E. His measurement was remarkably accurate, being less than 2% in error. To make this measurement, Eratosthenes used the relationship of alternate-interior angles of parallel lines.

As librarian of Alexandria, Eratosthenes had access to records of calendar events. He discovered that at noon on a certain day of the year, in a town on the Nile called Syene (now called Aswan), the sun was directly overhead. As a result, the bottom of a deep well was entirely lit and a vertical pole, being parallel to the rays hitting it, cast no shadow.

At the same time, however, a vertical pole in the city of Alexandria did cast a shadow. When that day arrived again, Eratosthenes measured the angle ($\angle 1$ in the figure below) formed by such a pole and the ray of light from the sun going past the top of the pole to the far end of the shadow. He found it to be about 7°12', or $\frac{1}{50}$ of 360°.

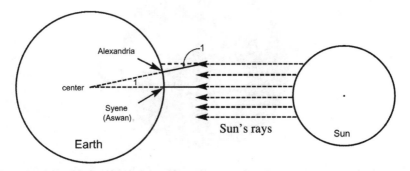

Assuming the rays of the sun to be parallel, he knew that the angle at the center of the earth must be congruent to $\angle 1$, and hence must also measure approximately $\frac{1}{50}$ of 360°. Since Syene and Alexandria were almost on the same meridian, Syene must be located on the radius of the circle, which was parallel to the rays of the sun. Eratosthenes thus deduced that the distance between Syene and Alexandria was $\frac{1}{50}$ of the circumference of the earth. The distance from Syene to Alexandria was believed to be about 5,000 Greek stadia. A *stadium* was a unit of measurement equal to the length of an Olympic or Egyptian stadium. Therefore, Eratosthenes concluded that the circumference of the earth was about 250,000 Greek stadia, or about 24,660 miles. This is very close to modern calculations. So how's that for some *real* geometry!

5.7. Surprising rope around the earth

This unit will show you that your intuition cannot always be trusted. This unit will surprise (or even shock) you. As always, take time to understand the situation and then try to grapple with it. Only then will the conclusion have its dramatic effect.

Consider the globe of the earth with a rope wrapped tightly around the equator. The rope will be about 24,900 miles long. We now lengthen the rope by exactly 1 yard. We position this (now loose) rope around the equator so that it is uniformly spaced off the globe. Will a mouse fit under the rope?

The traditional way to determine the distance between the circumferences is to find the difference between the radii. Let R be the length of the radius of the circle formed by the rope (circumference $C+1$) and r the length of the radius of the circle formed by the earth (circumference C).

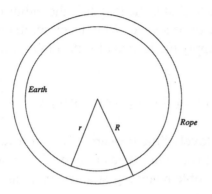

The familiar circumference formulas give us

$$C = 2\pi r, \text{ or } r = \frac{C}{2\pi}$$
and
$$C + 1 = 2\pi R, \text{ or } R = \frac{C+1}{2\pi}$$

We need to find the difference of the radii, which is

$$R - r = \frac{C+1}{2\pi} - \frac{C}{2\pi} = \frac{1}{2\pi} \approx .159 \text{ yards} \approx 5.7 \text{ inches}$$

Wow! There is a space of *more than $5\frac{1}{2}$ inches* for a mouse to crawl under.

You should really appreciate this astonishing result. Imagine, by lengthening the 24,900-mile rope by 1 yard, it lifted off the equator about $5\frac{1}{2}$ inches!

This unit lends itself to a very powerful problem-solving strategy that may be called "considering extreme cases."

Consider the original problem diagrammed above. You should realize that the solution was independent of the circumference of the earth, since the end result did not include the circumference in calculation. It only required calculating $\frac{1}{2\pi}$.

Here is a really nifty solution using an extreme case:

Suppose the inner circle (above) is very small, so small that it has a zero-length radius (that means it is actually just a point). We were required to find the difference between the radii, $R - r = R - 0 = R$.

So all we need to find is the length of the radius of the larger circle and our problem will be solved. With the circumference of the smaller circle now 0, we apply the formula for the circumference of the larger circle:

$$C + 1 = 0 + 1 = 2\pi R, \text{ then } R = \frac{1}{2\pi}$$

This unit has two lovely little treasures. First, it reveals an astonishing result, clearly not to be anticipated at the start, and, second, it provides you with a nice problem-solving strategy that can serve as a useful model for future use.

5.8. the lunes and the triangle

A lune is a crescent-shaped figure (such as that in which the moon often appears) formed by two circular arcs. The area of a circle is not typically commensurate with the areas of rectilinear figures.* A case in point is one of the so-called three famous problems of antiquity, namely, squaring the circle. That means we have now proved it impossible to construct a square (with unmarked straightedge and compasses) equal in area to a given circle. However, we shall provide you with a delightfully simple example where a circular area is equal to the area of a triangle.

Let's first recall the Pythagorean theorem. It states:

The sum of the squares of *the legs of a right triangle is equal to the square* of *the hypotenuse.*

This can be stated a bit differently with the same effect.

The sum of the squares on *the legs of a right triangle is equal to the square* on *the hypotenuse.*

---

*Arectlinear figure is one formed by straight lines.

We can take this a step further by stating:

The sum of the areas of *the squares on the legs of a right triangle is equal to* the area of *the square on the hypotenuse.*

As a matter of fact, we can easily show that the square can be replaced by any similar figures drawn on the sides of a right triangle. So we can state:

The sum of the areas of *the* similar polygons *on the legs of a right triangle is equal to* the area of *the* similar polygon *on the hypotenuse.*

This can then be restated for the specific case where the "similar polygons" are semicircles (which are, of course, similar):

The sum of the areas of *the semicircles on the legs of a right triangle is equal to* the area of *the semicircle on the hypotenuse.*

Thus, for the right triangle *ABC* below we can say that the area of the semicircles relates as follows:

Area *Q* + Area *R* = Area *P*

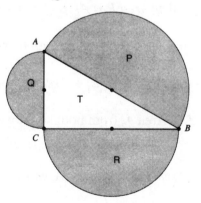

Suppose we now flip semicircle *P* over the rest of the figure (using $\overline{AB}$ as its axis). We would get a figure like the following.

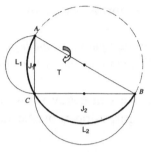

Let us now focus on the lunes formed by the two semicircles. We mark them L_1 and L_2.

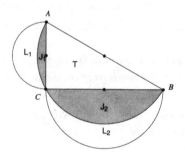

Earlier we established that *Area P = Area Q + Area R.*

In the figure above, that same relationship can be written as follows:

Area J$_1$ + Area J$_2$ + Area T = Area L$_1$ + Area J$_1$ + Area L$_2$ + Area J$_2$

If we subtract Area J_1 + Area J_2 from both sides, we get the following astonishing result:

$$Area\ T = Area\ L_1 + Area\ L_2$$

That is, we have a rectilinear figure (the triangle) equal in area to some non-rectilinear figures (the lunes). This is quite unusual since the measures of circular figures seem to always involve π, while rectangular (or straight line) figures do not.

5.9. the ever-present equilateral triangle

One of the most astonishing relationships in Euclidean geometry is a theorem first published by Frank Morley (writer Christopher Morley's father) in 1904. He discussed it with his colleagues at Cambridge University, yet he didn't publish it until 1924, while he was in Japan. To really appreciate the beauty of this theorem, you would be best off examining it with the Geometer's Sketchpad program. We will do the best we can to appreciate it here on these pages. The theorem states that

> *the adjacent angle trisectors* of any triangle intersect at three points, determining an equilateral triangle.*

Let us look at the following figure, where $\triangle ABC$ is just *any* triangle.

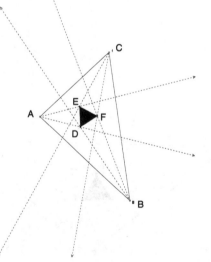

Notice how the points D, E, and F are the intersection points of the adjacent trisectors of the angles of $\triangle ABC$. And $\triangle DEF$ can be shown to be an equilateral triangle. Wow! This equilateral triangle evolved by

---

*Here we refer to the two angle trisectors (the rays that divide an angle into three equal parts) nearest a side of the triangle.

beginning with *any shaped* triangle. Geometer's Sketchpad allows you to change the shape of the original $\triangle ABC$ and observe that $\triangle DEF$ remains equilateral, although of different size.

Here are a few variations that you can create on Geometer's Sketchpad to witness this amazing relationship. This is truly one of the most dramatic (i.e., surprising) relationships in geometry and should be presented that way. Be cautioned, the proof is one of the most difficult in Euclidean geometry.*

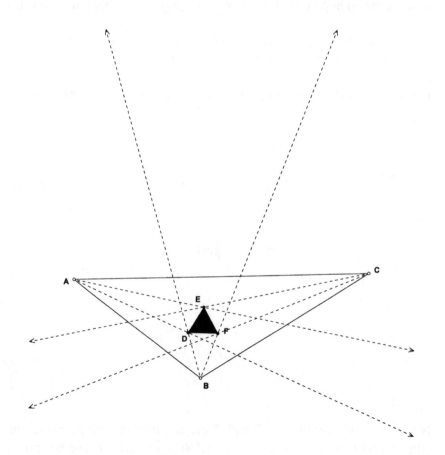

---

*Several proofs can be found in A. S. Posamentier and C. T. Salkind, *Challenging Problems in Geometry* (New York: Dover, 1996), pp. 158–61.

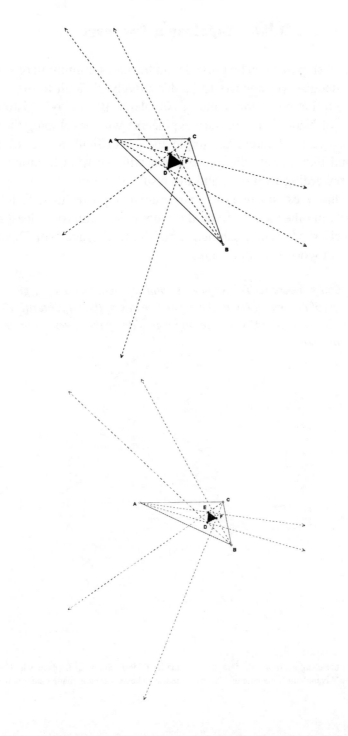

5.10. napoléon's theorem

One relationship that can be proved with the barest minimum of geometry knowledge—yet appears to be deceptively difficult to prove—is the theorem that bears the name of Napoléon. But today's historians credit one of Napoléon's military engineers* with developing the theorem. How can a theorem be simple and yet difficult to prove? This may sound like a contradiction, but you will see what it entails. It's actually rewarding to prove, and the result of the proof—that is, the theorem that is established—is extraordinarily powerful with lots of extensions. In other words, to do the proof can be fun (or at least generate a feeling of accomplishment), but the really nice "stuff" comes once we can work with the results.

Napoléon's theorem: *The segments joining each vertex of a given triangle (of any shape) with the remote vertex of the equilateral triangle, drawn externally on the opposite side of the given triangle, are congruent.*

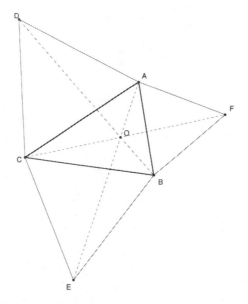

---

*One of these engineers was the famous mathematician Pierre-Simon de Laplace, who first met sixteen-year-old Napoléon Bonaparte in 1785 when he was to be examined in mathematics at the military academy.

That is, when $\triangle ADC$, $\triangle BCE$, and $\triangle ABF$ are equilateral, $\overline{AE}$, $\overline{BD}$, and $\overline{CF}$ are congruent to one another. Take note of the unusual nature of this situation, since we started with *any* triangle and still this relationship holds true. If you were to draw your own original triangle, you would come up with the same conclusion. Either straightedge and compasses or Geometer's Sketchpad would be fine for this; however, the latter would be better.

Before we embark on the adventures that this theorem holds, it may be helpful to give you a hint as to how to prove this theorem. The trick is to identify the proper triangles to prove congruent. They are not easy to identify. One pair of these triangles is shown below. These two congruent triangles will establish the congruence of $\overline{AE}$ and $\overline{BD}$. The other segments can be proved congruent in a similar way with another pair of congruent triangles, embedded in the figure as these are.

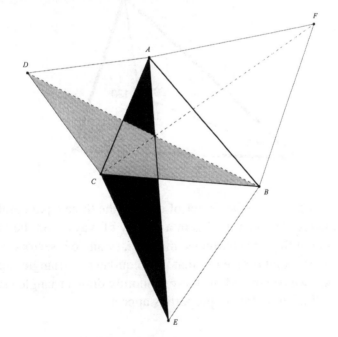

There are quite a few unusual properties in this figure. For example, you may not have noticed that the three segments $\overline{AE}$, $\overline{BD}$, and $\overline{CF}$ are also concurrent (i.e., contain a common point). This concept is not explored much in the typical high school geometry course. Yet, it's not to be taken for granted. It must be proved, but for our purposes we shall accept it without proof.*

Not only is point O a common point for the three segments, but it is also the only point in the triangle where the sum of the distances to the vertices of the original triangle is a minimum. This is often called the *minimum-distance point* of the triangle *ABC*. The sum of the distances from any other point in the triangle to the three vertices would be greater than from this minimum-distance point.

As if this weren't enough, this point, O, is the only point in the triangle where the sides subtend equal angles. That is, $m\angle AOC = m\angle COB = m\angle BOA = 120°$.

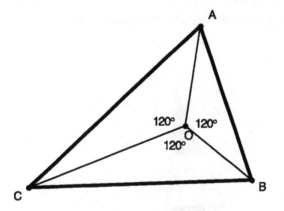

There is more! Locate the center of each of the three equilateral triangles following. You can do this in a variety of ways: find the point of intersection of the three altitudes, medians, or angle bisectors. Joining these center points reveals that an equilateral triangle appears. Remember, we began with just any randomly drawn triangle ($\triangle ABC$) and now all of these lovely properties appear.

---

*For a proof of this theorem and its extensions, see A. S. Posamentier, *Advanced Euclidean Geometry: Excursions for Secondary Teachers and Students* (Emeryville, Calif.: Key College Publications, 2002), chap. 4.

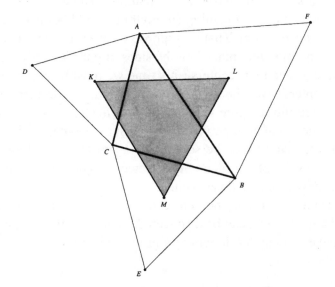

With a computer geometry program, such as Geometer's Sketchpad, you can see that, regardless of the shape of the original triangle, the above relationships all hold true. An interesting question you might ask yourself is: What would you expect to happen if point C were to be on $\overline{AB}$, thereby collapsing the original triangle? See the figure below.

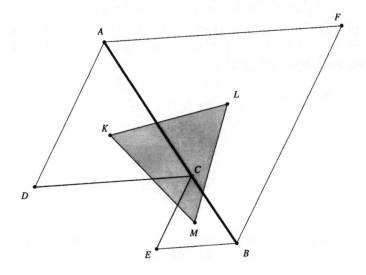

Lo and behold, we still have our equilateral triangle *KLM* preserved. Perhaps even more astonishing (if anything could be) is the generalization of this theorem. That is, suppose we were to construct similar triangles appropriately placed on the sides of our randomly drawn triangle, and joined their centers (this time we must be consistent as to which "centers" we choose to use—centroid, orthocenter, incenter, etc.).* The resulting figure will be similar to the three similar triangles.

With the aid of geometry software such as Geometer's Sketchpad, you can see that all that we said above about triangles drawn *externally* on the sides of our randomly selected triangle can be extended to triangles drawn *internally* as well.

So from this simple theorem came a host of incredible relationships, many of which can be discovered independently—in particular, if you have a computer drawing program available.

5.11. the golden rectangle

When we talk about the beauty of mathematics, we could talk about that which most artists think is the most beautiful rectangle. This rectangle, often called the Golden Rectangle, has been shown by psychologists to be the most esthetically pleasing rectangle compared to any other rectangle. It is often used in architecture and art. For example, the Parthenon in Athens, Greece, is based on the shape of a Golden Rectangle. If we outline many figures in classical art, the Golden Rectangle will predominate.

---

*The *centroid* is the point of intersection of the medians of a triangle. The *orthocenter* is the point of intersection of the altitudes of a triangle. The *incenter* is the point of intersection of the angle bisectors of a triangle.

To construct a Golden Rectangle, begin with a square $ABEF$ (see the top figure on p. 176). Locate the midpoint M of one side of the square $ABEF$ and make a circular arc with center at M and radius length ME. Call the point D where the arc intersects $\overrightarrow{AF}$. Then erect a perpendicular to $\overline{AD}$ at D to meet $\overrightarrow{BE}$ at C. Rectangle $ABCD$ is a Golden Rectangle. This can be done with straightedge and compasses, or you may now be adept at using Geometer's Sketchpad.

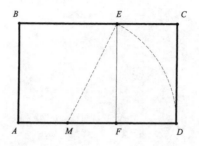

Here is a nice extension of the Golden Rectangle. If we continuously construct squares in the Golden Rectangle, as shown below, each resulting rectangle is a Golden Rectangle; that is, it is similar to the original rectangle, since all Golden Rectangles are similar.

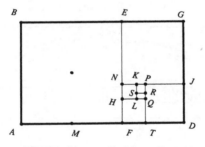

Remove square *ABEF* from Golden Rectangle *ABGD* to get Golden Rectangle *EGDF*.

Remove square *EGJN* from Golden Rectangle *EGDF* to get Golden Rectangle *JDFN*.

Remove square *PJDT* from Golden Rectangle *JDFN* to get Golden Rectangle *TFNP*.

Remove square *TFHQ* from Golden Rectangle *TFNP* to get Golden Rectangle *HNPQ*.

Remove square *HNKL* from Golden Rectangle *HNPQ* to get Golden Rectangle *KPQL*.

Remove square *KPRS* from Golden Rectangle *KPQL* to get Golden Rectangle *SRQL*.

And so on.

Notice that each time a square is taken from a Golden Rectangle the resulting rectangle is also a Golden Rectangle.

Once you have drawn the above figure, you ought to draw quarter circular arcs as shown below. The resulting figure approximates a logarithmic spiral.

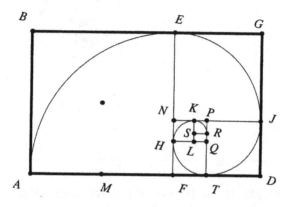

We can locate the vanishing point of the spiral by drawing the diagonals of the two largest Golden Rectangles as shown in the figure below.

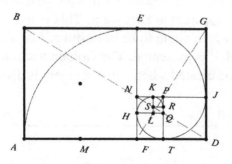

If you construct this figure accurately, you will see that $\overline{BD}$ and $\overline{GF}$ contain the diagonals of the other Golden Rectangles as well. Moreover, $\overline{BD} \perp \overline{GF}$.

A similar spiral can be drawn by locating the centers of each of the squares in succession from largest to smallest and drawing the following:

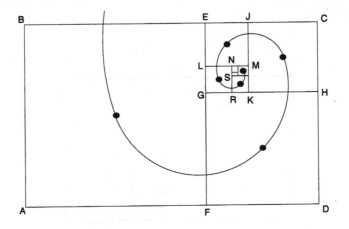

There is no end to the beauty of this rectangle! Hence it is called the *Golden Rectangle*.

5.12. the golden Section Constructed by paper folding

There are many things in mathematics that are beautiful, yet sometimes the beauty is not apparent at first sight. This is not the case with the Golden Section, which ought to be beautiful at first sight regardless of the form in which it is presented. The Golden Section refers to a proportion that can be created when a line segment is divided by a point.

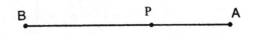

Simply, for the segment $\overline{AB}$, the point P partitions (or divides) it into two segments, $\overline{AP}$ and $\overline{BP}$, such that $\frac{AP}{PB} = \frac{PB}{AB}$. This proportion, apparently already known to the Egyptians and the Greeks, was probably first named the Golden Section or *sectio aurea* by Leonardo da

Vinci, who drew geometric diagrams for Fra Luca Pacioli's book, *De Divina Proportione* (1509), which dealt with this topic.

There are probably endless beauties involving this Golden Section. One of these is the relative ease with which one can construct the ratio by merely folding a strip of paper.

Simply take a strip of paper, say about 1″ or 2″ wide, and make a knot. Then very carefully flatten the knot as shown in the figure below. Notice the resulting shape appears to be a regular pentagon, that is, a pentagon with all angles congruent and all sides the same length.

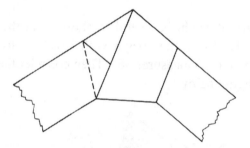

If you use relatively thin translucent paper and hold it up to a light, you ought to be able to see a pentagon with its diagonals.* These diagonals intersect each other in the Golden Section (see below).

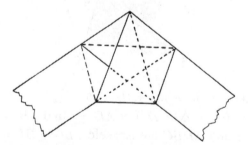

Let's take a closer look at the following pentagon. Point D divides $\overline{AC}$ into the Golden Section, since $\frac{DC}{AD} = \frac{AD}{AC}$.

We can say that the segment of length AD is the mean proportional between the lengths of the shorter segment ($\overline{DC}$) and the entire segment ($\overline{AC}$).

---

*These diagonals form the regular five-cornered star called a regular pentagram.

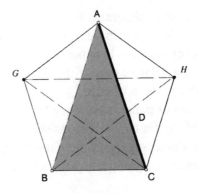

For some readers, it might be useful to show what the value of the Golden Section is. To do this, begin with isosceles triangle ABC,* whose vertex angle ($\angle A$) measures 36°. Then consider the bisector $\overline{BD}$ of $\angle ABC$ (see figure below):

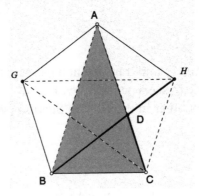

We find that $m\angle DBC = m\angle ABD = 36°$.

Therefore $\triangle ABC \sim \triangle BCD$. Let $AD = x$ and $AB = 1$.

Since $\triangle ADB$ and $\triangle DBC$ are isosceles, $BC = BD = AD = x$.

From the similarity above: $\frac{1-x}{x} = \frac{x}{1}$.

This gives us: $x^2 + x - 1 = 0$, and $x = \frac{\sqrt{5}-1}{2}$ (The negative root cannot be used for the length of $\overline{AD}$.)

We recall (see section 4.8) that $\frac{\sqrt{5}-1}{2} = \frac{1}{\phi}$.

The ratio for $\triangle ABC$ of $\frac{side}{base} = \frac{1}{x} = \phi$, which is the Golden Ratio.

We therefore call this a *Golden Triangle*.

---

*An *isosceles triangle* is one that has two sides of equal length.

5.13. the regular pentagon that isn't

One of the more difficult constructions to do using an unmarked straightedge and compasses is that of the regular pentagon. There are many ways to do this construction, yet none are particularly easy. You might try to develop a construction on your own, realizing that the Golden Section is involved here.

For years engineers have been using a method for drawing what appears to be a regular pentagon, yet careful inspection will show that the construction is a tiny bit irregular.* This method, which is provided below, was developed in 1525 by the famous German artist Albrecht Dürer.

We refer to the diagram below. Beginning with a segment AB, five circles of radius AB are constructed as follows:

1. Draw circles, with centers at A and B, to intersect at Q and N.
2. Then the circle with center Q is drawn to intersect circles A and B at points R and S, respectively.
3. $\overline{QN}$ intersects circle Q at P.
4. $\overrightarrow{SP}$ and $\overrightarrow{RP}$ intersect circles A and B at points E and C, respectively.
5. Draw the circles with centers at E and C, with radius AB to intersect at D.
6. The polygon $ABCDE$ is (supposedly) a regular pentagon.

---

*For a discussion of where the error lies, see, Alfred S. Posamentier and Herbert A. Hauptman, *101 Great Ideas for Introducing Key Concepts in Mathematics* (Thousand Oaks, Calif.: Corwin Press, 2001), pp. 141–46.

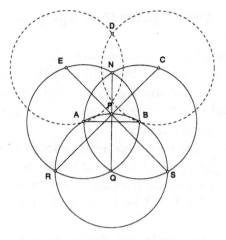

Joining the points in order, we get the pentagon *ABCDE*.

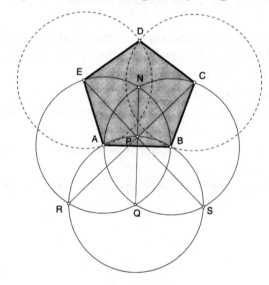

Although the pentagon "looks" regular, $\angle ABC$ is about $\frac{22}{60}$ of a degree too large. That is, for $ABCDE$ to be a regular pentagon, each angle must measure 108°; instead we have $m\angle ABC \approx 108.3661202°$. You might try to draw this with Geometer's Sketchpad or simply on a piece of paper. It ought to be easy to draw this artwork, following the instructions provided.

5.14. Pappus's Invariant

One of the lovely relationships in geometry occurs when something remains true regardless of the shape of the figure. That is, we can draw something from instructions given over the telephone, where the appearance of the figure drawn will vary with each individual, but one part of it will be common to all drawings. We call this an *invariant.* Such a situation has been handed down to us by Pappus of Alexandria (ca. 300–350 C.E.) from his *Collection,** a compilation of most of what was known in geometry at that time. Let us look at what he presents and just marvel at it.

Consider any two lines, each with three points located anywhere on the lines. Then connect the points of the first line to those on the second line, but keep from connecting the corresponding points. That is, don't connect the right-most point on one line to the right-most point on the other line, or don't connect the two middle points.

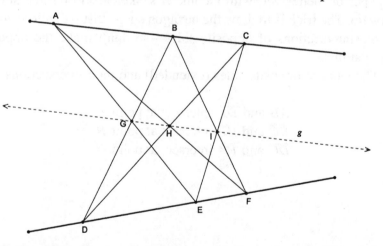

In the above figure, three points of intersection *G, H,* and *I* are marked. Now here is the amazing part: regardless of how you drew the original lines or where you located the points on the lines, the points *G, H,* and *I* are always collinear (that is, they lie on the same straight line)!†

---

*It is listed in the *Collection* as Lemma 13, proposition 139.

†For a proof of Pappus's theorem see Alfred S. Posamentier, *Advanced Euclidean Geometry: Excursions for Secondary Teachers and Students* (Emeryville, Calif.: Key College Press, 2002).

This astonishing result is just another piece of evidence of the beauty of geometry. You might want to draw a few other variations of the above description to firm up this amazing result.

5.15. Pascal's Invariant

In 1640, at the age of sixteen, the famous mathematician Blaise Pascal published a one-page paper entitled *Essay pour les coniques*, which presents us with a most insightful theorem. What he called *mysterium hexagrammicum* states that *the intersections of the opposite sides of a hexagon inscribed in a conic section* are collinear*† (that is, they lie on the same straight line). To make our discussion as simple as possible, we shall use the most common conic section: a circle.

Consider the hexagon *ABCDEF* inscribed in the circle (that is, all its vertices are on the circle). You might try this independently, either on paper or, better yet, with Geometer's Sketchpad or some similar software. The trick is to draw the hexagon shape that will allow you to get the intersections of opposite sides—so don't make the opposite sides parallel.

The pairs of opposite sides (extended) and their intersections are

$\overline{AB}$ and $\overline{DE}$ intersect at point I
$\overline{BC}$ and $\overline{EF}$ intersect at point H
$\overline{DC}$ and $\overline{FA}$ intersect at point G

---

*Aconic section is a curve that results from a plane cutting a cone. These include a circle, an ellipse, a hyperbola, and a parabola.

†For a proof of Pascal's theorem, see Alfred S. Posamentier, *Advanced Euclidean Geometry: Excursions for Secondary Teachers and Students* (Emeryville, Calif.: Key College Press, 2002).

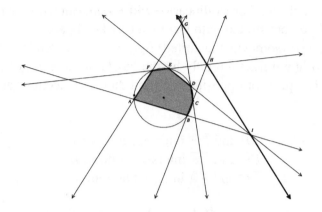

Notice points *G*, *H*, and *I* lie on the same straight line.

Here is a differently shaped hexagon inscribed in a circle. Again notice that regardless of the shape, the three points of intersection of the opposite sides of the hexagon lie on a straight line (i.e., the points *G*, *H*, and *I* are collinear).

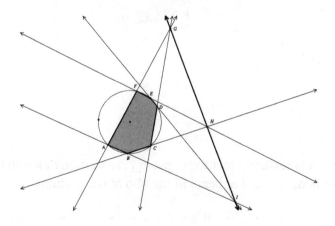

If you do it on a computer, you can actually see that, despite changing the shape of the hexagon, the collinearity of points *G*, *H*, and *I* always remains intact. The amazing thing about this situation is that it is independent of the shape of the hexagon. You can even distort the hexagon so that it doesn't look like a polygon anymore, and as long as you keep track of what were the opposite sides, and that the vertices remain on a circle, then collinearity will remain intact. Again, this is very easily demonstrated with Geometer's Sketchpad.

In the next (somewhat distorted) figure you can identify what used to be the original hexagon only by the labels of the sides of the original one. The pairs of opposite sides and their intersections are just as before:

$$\overline{AB} \text{ and } \overline{DE} \text{ intersect at point } I$$
$$\overline{BC} \text{ and } \overline{EF} \text{ intersect at point } H$$
$$\overline{DC} \text{ and } \overline{FA} \text{ intersect at point } G$$

These intersection points, *G*, *H*, and *I*, are still collinear.

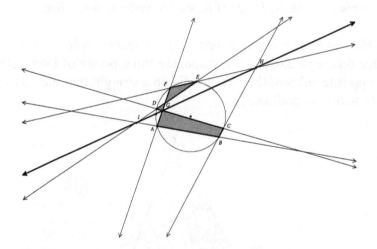

This truly amazing relationship can give us motivation to look into why this behaves as it appears in the above illustrations.

5.16. brianchon's ingenius extension of pascal's idea

In 1806, at the age of twenty-one, a student at the École Polytechnique, Charles Julien Brianchon (1785–1864) published an article in the *Journal de L'École Polytechnique* that was to become one of the fundamental contributions to the study of conic sections in projective geometry.* His development led to a restatement of the somewhat forgotten theorem of Pascal and its extension, after which Brianchon stated a new theorem, which later bore his name. Brianchon's theorem,† which states, "In any hexagon circumscribed about a conic section, the three diagonals cross each other in the same point,"‡ bears a curious resemblance to Pascal's theorem.

To fully appreciate the relationship between Pascal's theorem and what Brianchon discovered, it is best to first understand what the concept of *duality* in mathematics is. Two statements are *duals* of one another when all of the key words in the statements are replaced by their *dual* words. For example, *point* and *line* are dual words, *collinearity* and *concurrency* are duals, *inscribed* and *circumscribed* are duals, *sides* and *vertices* are duals, and so on. Here is an example of the duality relationship. Notice how the terms *point* and *line* have been interchanged.

> Two *points* determine a *line*.
> Two *lines* (intersecting, of course) determine a *point*.

Below you will see Pascal's theorem restated and next to it Brianchon's theorem. Notice that the italicized words in Pascal's statement are replaced by their duals, forming Brianchon's statement. Thus, they are, in fact, duals of one another.

Pascal's theorem	**Brianchon's theorem**
The *points of intersection* of the opposite *sides* of a hexagon *inscribed in* a conic section are *collinear*.	The *lines joining* the opposite *vertices* of a hexagon *circumscribed about* a conic section are *concurrent*.

In the figure below, the hexagon *ABCDEF* is circumscribed about the circle. As with Pascal's theorem before, we shall consider only a

---

*Projective geometry is the study of geometric properties that are invariant under projection.

†For a proof of Brianchon's theorem, see Alfred S. Posamentier, *Advanced Euclidean Geometry: Excursions for Secondary Teachers and Students* (Emeryville, Calif.: Key College Press, 2002).

‡D. E. Smith, ed., *Source Book in Mathematics*, (New York: McGraw-Hill, 1929), p. 336.

simple conic section, that is, a circle. According to Brianchon's statement, the lines containing opposite vertices are concurrent. You can easily experiment with differently shaped circumscribed hexagons to verify that it is true. Again, we see that the simplicity of this figure and its result makes for its beauty.

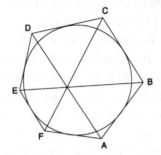

Right after stating his theorem, he suggested that if points *A*, *F*, and *E* were to be moved so that they would be collinear, with vertex *F* becoming a point of tangency, and thereby forming a pentagon, the same statement could be made. That is, since pentagon *ABCDE* is circumscribed about a circle, then $\overline{CF}$, $\overline{AD}$, and $\overline{BE}$ are concurrent.

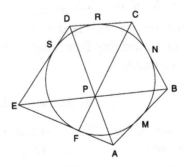

Take a moment to marvel about the hexagon situation and allow yourself to appreciate its connection to the pentagon above. This should really entice you to further investigations!

5.17. a Simple proof of the pythagorean theorem

One of the most celebrated relationships in mathematics is the Pythagorean theorem. Why is this so much in the minds of adults, who usually remember this above all else learned in school mathematics? Could this be because we usually refer to the theorem with the first three letters of the alphabet, $a^2 + b^2 = c^2$, and it is like learning your ABCs?

Clearly, the Pythagorean theorem is the basis for much of geometry and all of trigonometry. For this reason one must be careful about discovering a new proof to make sure that it is not based on a relationship established by the Pythagorean theorem itself. Such is the case with trigonometry. No proof of the Pythagorean theorem can use trigonometric relationships, since they are based on the Pythagorean theorem and were we to do this, it would be a clear case of circular reasoning—not legitimate!

Now to the proof. It is very simple, but it is based on a theorem which states that *when two chords intersect in a circle, the product of the segments of one chord is equal to the product of the segments of the other chord.*

In the figure below, this would mean that for the two intersecting chords $p \cdot q = r \cdot s$

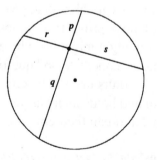

Now let us consider the circle with diameter $\overline{AB}$ perpendicular to chord $\overline{CD}$.

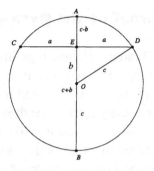

This time the product of the segments of the two intersecting chords gives us $(c - b)(c + b) = a^2$.

Then, $c^2 - b^2 = a^2$, and therefore, $a^2 + b^2 = c^2$. The Pythagorean theorem is proved again. Although there have been many other proofs after its publication, a nice collection of 370 proofs of the Pythagorean theorem was published under the title *The Pythagorean Proposition*, by Elisha S. Loomis in 1940 and republished by the National Council of Teachers of Mathematics in 1968.

5.18. folding the pythagorean theorem

After all the struggles students go through to prove the Pythagorean theorem, imagine, we will now prove this famous theorem by simply folding paper. Your first thought might be, why didn't my teachers ever show me this when I was in school? Agood question. Perhaps that is one of the reasons why many adults need to be convinced in later life that mathematics is beautiful and holds many as yet unexposed delights.

We can take the Pythagorean theorem, which states that

the sum of the squares* *on the legs of a right triangle is equal to the* square *on the hypotenuse of the triangle,*

and propose that

---

*We refer to the *area* of the square, as we will refer to the *area* of the right triangles in the following statement.

the sum of the areas of similar polygons *on the legs of a right triangle is equal to the area of the* similar polygon *on the hypotenuse
of the triangle.*

Consider the following right triangle with altitude $\overline{CD}$:

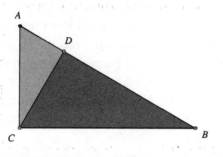

Notice that $\triangle ADC \sim \triangle CDB \sim \triangle ACB$. In the above figure, $\triangle ADC$ and
$\triangle CDB$ are folded over (along $\overline{AC}$ and $\overline{CB}$, respectively) onto
$\triangle ACB$. So clearly $Area\triangle ADC + Area\triangle CDB = Area\triangle ACB$. If we
unfold the triangles (including the folded-over $\triangle ACB$ itself), we get the
following that shows that the relationship of the similar polygons (here
right triangles) is the extension of the Pythagorean theorem: *the sum of
the areas of similar* right triangles *on the legs of a right triangle is equal
to the area of the similar* right triangle *on the hypotenuse of the triangle.*

This essentially "proves" the Pythagorean theorem by paper
folding! See below.

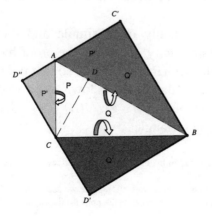

5.19. President James A. Garfield's Contribution to Mathematics

We shall begin by asking the famous question: What do these three men have in common: Pythagoras, Euclid, and James A. Garfield (1831–1881), the twentieth president of the United States?

After some moments of perplexity, I shall relieve your frustration by telling you that all three fellows proved the Pythagorean theorem. The first two fellows should not be much of a surprise, but President Garfield? He wasn't a mathematician. He didn't even study mathematics. As a matter of fact, his only study of geometry, some twenty-five years before he published his proof of the Pythagorean theorem, was informal and alone.*

While a member of the House of Representatives, Garfield, who enjoyed "playing" with elementary mathematics, stumbled upon a cute proof of this famous theorem. It was subsequently published in the *New England Journal of Education* after Garfield was encouraged by two professors (Quimby and Parker) at Dartmouth College, where he went to give a lecture on March 7, 1876. The text begins with the following:

> In a personal interview with General James A. Garfield, Member of Congress from Ohio, we were shown the following demonstration of the pons asinorum,† which he had hit upon in some mathematical amusements and discussions with other M.C.'s. We do not remember to have seen it before, and we think it something on which the members of both houses can unite without distinction of party.

Garfield's proof is actually quite simple and therefore, beautiful. Beauty is often in simplicity! We begin the proof by placing two congruent right triangles ($\triangle ABE \cong \triangle DCE$) so that points B, C, and E are collinear, as shown in the figure below, and a trapezoid is formed. Notice also that since $m\angle AEB + m\angle CED = 90°$, then $m\angle AED = 90°$, making $\triangle AED$ a right triangle.

---

*In October 1851 he noted in his diary that "I have today commenced the study of geometry alone without class or teacher."

†This would appear to be a wrong reference, since we usually consider the proof that the base angles of an isosceles triangle are congruent as the pons asinorum, or "bridge of fools."

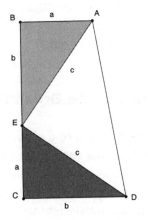

The area of the trapezoid $= \frac{1}{2}$(sum of bases)(altitude)

$$= \frac{1}{2}(a+b)(a+b)$$

$$= \frac{1}{2}a^2 + ab + \frac{1}{2}b^2$$

The sum of the areas of the three triangles (also the area of the trapezoid) is

$$= \frac{1}{2}ab + \frac{1}{2}ab + \frac{1}{2}c^2$$

$$= ab + \frac{1}{2}c^2.$$

We now equate the two expressions for the area of the trapezoid

$$\frac{1}{2}a^2 + ab + \frac{1}{2}b^2 = ab + \frac{1}{2}c^2$$

$$\frac{1}{2}a^2 + \frac{1}{2}b^2 = \frac{1}{2}c^2$$

which is the familiar $a^2 + b^2 = c^2$, also known as the *Pythagorean theorem*.

There are more than four hundred proofs* of the Pythagorean theorem available today; many are ingenious, yet some are a bit cumbersome. However, none will ever use trigonometry. Why is this? There can be no proof of the Pythagorean theorem using trigonometry, since trigonometry depends (or is based) on the Pythagorean theorem. Thus,

---

*A classic source for 370 proofs of the Pythagorean theorem is Elisha S. Loomis's *The Pythagorean Proposition* (Reston, Va.: NCTM, 1968).

using trigonometry to prove the very theorem on which it depends would be circular reasoning. Have you been motivated to discover a new proof of this most famous theorem?

5.20. What is the area of a Circle?

You have been "told" that the area of a circle is found by the formula $A = \pi r^2$. Most likely you have not been given an opportunity to discover where this formula may have come from or how it relates to other concepts. It is not only entertaining, but also instructionally sound, to have the formula evolve from previously learned concepts. Recall the formula for finding the area of a parallelogram (i.e., the product of its base and height). Fortified with this basic formula, we will present a nice justification for the formula for the area of a circle.

Begin by drawing a convenient size circle on a piece of cardboard. Divide the circle into 16 equal arcs. This may be done by marking off consecutive arcs of 22.5° or by consecutively dividing the circle into two equal parts, then four equal parts, then bisecting each of these quarter arcs, and so on.

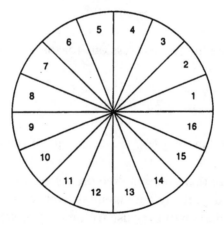

These sectors, shown above, are then to be cut apart and placed in the manner shown in the figure below.

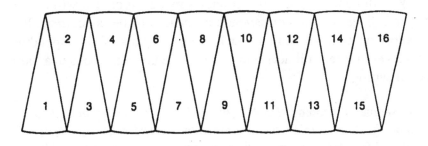

This placement suggests that we have a figure that "approximates" a parallelogram. That is, were the circle cut into more equal sectors, then the figure would look even more like a true parallelogram, without the small circular arcs comprising the sides. Let us assume it is a parallelogram. In this case, the base would have length $\frac{1}{2}C$, where $C = 2\pi r$ (r is the radius and C is the circumference). The area of the parallelogram is equal to the product of its base and altitude (which here is r). Therefore, the area of the parallelogram $= (\frac{1}{2}C)(r) = \frac{1}{2}(2\pi r)(r) = \pi r^2$, which is the commonly known formula for the area of a circle.

5.21. a Unique placement of two triangles

Most of the geometry that we study in school is not dependent on the placement of the figures. That is, they can be placed anywhere on the plane (a sheet of paper). Usually, where they are placed in relation to other figures is not considered. However, there is a very important position relationship that we will inspect. This relationship actually forms the basis for a branch of geometry called projective geometry, which was discovered in 1648 by Gérard Desargues (1591–1661).

We are going to consider two triangles, whose "corresponding vertices"—which we will designate with the same letter—will determine the corresponding sides. This is important to keep in mind as we move along. The two triangles are going to be situated in a very specific manner, and their shape (or relative shape) is of no real concern to us. This is quite different from the kind of thinking used in the high school study of geometry.

We will be placing any two triangles in a position that will enable the three lines joining corresponding vertices to be concurrent. Remarkably enough, when this is achieved, the pairs of corresponding sides meet in three collinear points. Let's see how this looks in a more formal setting.

> Desargues's theorem: *If $\triangle A_1 B_1 C_1$ and $\triangle A_2 B_2 C_2$ are situated so that the lines joining the corresponding vertices, $\overleftrightarrow{A_1 A_2}$, $\overleftrightarrow{B_1 B_2}$, and $\overleftrightarrow{C_1 C_2}$, are concurrent, then the pairs of corresponding sides intersect in three collinear points.*

In the figure, the lines joining the corresponding vertices, $\overleftrightarrow{A_1 A_2}$, $\overleftrightarrow{B_1 B_2}$, and $\overleftrightarrow{C_1 C_2}$, all meet at P.

The extensions of the corresponding sides meet at points A', B', and C' as follows:

> Lines $\overleftrightarrow{B_2 C_2}$ and $\overleftrightarrow{B_1 C_1}$ meet at A';
> lines $\overleftrightarrow{A_2 C_2}$ and $\overleftrightarrow{A_1 C_1}$ meet at B'; and
> lines $\overleftrightarrow{B_2 A_2}$ and $\overleftrightarrow{B_1 A_1}$ meet at C'.

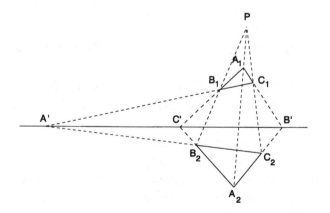

This is truly remarkable, but to make it even more astonishing, the converse is also true. Namely, *if $\triangle A_1 B_1 C_1$ and $\triangle A_2 B_2 C_2$ are situated so that the pairs of corresponding sides intersect in three collinear points, then the lines joining the corresponding vertices, $\overleftrightarrow{A_1 A_2}$, $\overleftrightarrow{B_1 B_2}$, and $\overleftrightarrow{C_1 C_2}$ are concurrent.* For the teacher who wishes to pursue this theorem further, it is useful to know that it is a self-dual. That is, the dual of the theorem is the converse.*

5.22. a point of invariant distance in an equilateral triangle

Equilateral triangles are the most symmetric triangles. The angle bisectors, the altitudes, and the medians are all the same line segments. No other triangle can boast this property. Their point of intersection is the center of the inscribed and circumscribed circles, again a unique property. These ought to be well-known properties. What is not well known is that if any point is chosen in an equilateral triangle, the sum of the perpendicular distances to the sides of the triangle is constant. As a matter of fact, this sum is equal to the length of the altitude of the triangle.

A very elegant (or somewhat sophisticated) method for verifying this is by taking an "extreme" point. By taking the "any point" to be a vertex, this can be easily established. Then the sum of the distances to two of the sides is zero, leaving the distance to the third side as the sum. This distance to the third side is simply the altitude.

We can show this in a number of more traditional ways.

We seek to prove *the sum of the distances from any point in the interior of an equilateral triangle to the sides of the triangle is constant (the length of the altitude of the triangle).*

Here is an illustration of what is being established.

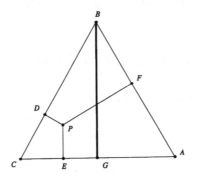

---

*The proof of the theorem can be found in Alfred S. Posamentier, *Advanced Euclidean Geometry: Excursions for Secondary Teachers and Students* (Emeryville, Calif.: Key College, 2002).

Two proofs of this interesting property are provided here. The first compares the length of each perpendicular segment to a portion of the altitude, and the second involves area comparisons.

Proof 1: In equilateral $\triangle ABC$ below, $\overline{PR} \perp \overline{AC}$, $\overline{PQ} \perp \overline{BC}$, $\overline{PS} \perp \overline{AB}$, and $\overline{AD} \perp \overline{BC}$. Draw a line through P parallel to $\overline{BC}$, meeting $\overline{AD}$, $\overline{AB}$, and $\overline{AC}$ at G, E, and F, respectively.

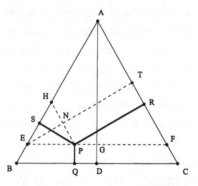

Since $PGDQ$ is a rectangle, $PQ = GD$. Draw $\overline{ET} \perp \overline{AC}$.

Since $\triangle AEF$ is equilateral, $\overline{AG} \cong \overline{ET}$ (all the altitudes of an equilateral triangle are congruent).

Draw $\overline{PH} \parallel \overline{AC}$, meeting $\overline{ET}$ at N. $\overline{NT} \cong \overline{PR}$.

Since $\triangle EHP$ is equilateral, altitudes $\overline{PS}$ and $\overline{EN}$ are congruent.

Therefore, we have shown that $PS + PR = ET = AG$.

Since $PQ = GD, PS + PR + PQ = AG + GD = AD$, a constant for the given triangle.

Proof 2: In equilateral $\triangle ABC$, $\overline{PR} \perp \overline{AC}$, $\overline{PQ} \perp \overline{BC}$, $\overline{PS} \perp \overline{AB}$, and $\overline{AD} \perp \overline{BC}$.

Draw $\overline{PA}$, $\overline{PB}$, and $\overline{PC}$ as follows.

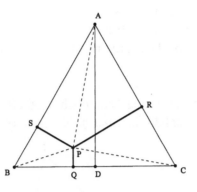

The area of $\triangle ABC$ = area of $\triangle APB$ + area of $\triangle BPC$ + area of $\triangle CPA$
  = $\frac{1}{2}(AB)(PS) + \frac{1}{2}(BC)(PQ) + \frac{1}{2}(AC)(PR)$.
Since $AB = BC = AC$, the area of $\triangle ABC = \frac{1}{2}(BC)[PS + PQ + PR]$.
However, the area of $\triangle ABC = \frac{1}{2}(BC)(AD)$.
Therefore, $PS + PQ + PR = AD$, a constant for the given triangle.

Take a look at what was just proved, that from any point in the triangle the sum of the distances to the sides is constant. Suppose the point at issue were on a side of the triangle; would the same property hold true? This is just one of the questions that can be raised to further embellish this new concept.

5.23. the nine-Point Circle

Perhaps one of the true joys in geometry is to observe how some seemingly unrelated points are truly related to each other. We begin with the very important notion that any three non-collinear points determine a circle. That is, whenever one has three non-collinear points, there is always a unique circle that contains the three points. When a fourth point also emerges on the same circle, it is quite noteworthy. Imagine that we can show that nine points all end up being on the same circle. That is phenomenal! These nine points, for any given triangle, are

- the midpoints of the sides,
- the feet of the altitudes,* and
- the midpoints of the segments from the orthocenter† to the vertices.

They have the surprising relationship of all being on the same circle. This circle is called the *nine-point circle* of the triangle.

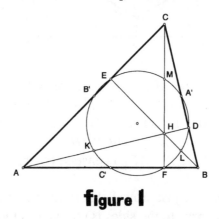

figure 1

In 1765 Leonhard Euler showed that six of these points, the midpoints of the sides and the feet of the altitudes, determine a unique circle. Yet not until 1820, when a paper‡ published by Charles Julien Brianchon and Jean Victor Poncelet appeared, were the remaining three points (the midpoints of the segments from the orthocenter to the vertices) found to be on this circle. The paper contains the first complete proof of the theorem and uses the name "the nine-point circle" for the first time.

> The nine-point circle theorem: *In any triangle, the midpoints of the sides, the feet of the altitudes, and the midpoints of the segments from the orthocenter to the vertices lie on a circle.*

Proof: To simplify the discussion of this proof, we shall consider each part with a separate diagram. Bear in mind, though, that each of

---

*The feet of the altitudes refer to the points at which the altitudes intersect the sides.
†The orthocenter is the point of intersection of the altitudes of a triangle.
‡*Recherches sur la determination d'une hyperbole équilatèau moyen de quartes conditions données* (Paris, 1820).

figures 2–5 is merely an extraction from figure 1 (above), which is the complete diagram. In other word the following development builds up to the fifth figure in small increments.

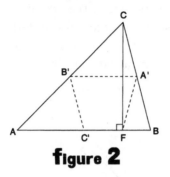

figure 2

In the figure above, points A', B', and C' are the midpoints of the three sides of $\triangle ABC$ opposite their respective vertices. $\overline{CF}$ is an altitude of $\triangle ABC$. Since $\overline{A'B'}$ is a midline of $\triangle ABC$, $\overline{A'B'} \parallel \overline{AB}$. Therefore, quadrilateral $A'B'C'F$ is a trapezoid. $B'C'$ is also a midline of $\triangle ABC$, so that $B'C' = \frac{1}{2}BC$. Since $\overline{A'F}$ is the median to the hypotenuse of right $\triangle BCF$, $A'F = \frac{1}{2}BC$. Therefore, $B'C' = A'F$ and trapezoid $A'B'C'F$ is isosceles.

You will recall that when the opposite angles of a quadrilateral are supplementary,* as in the case of an isosceles trapezoid, the quadrilateral is cyclic.† Therefore quadrilateral $A'B'C'F$ is cyclic.

So far, we have four of the nine points on one circle.

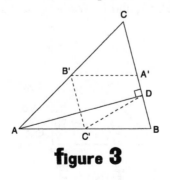

figure 3

---

*Two angles are supplementary if their measures have a sum of 180°.

†Acyclic quadrilateral is one whose four vertices lie on the same circle. It can also be called an inscribed quadrilateral.

To avoid any confusion, we redraw $\triangle ABC$ (above) and include altitude $\overline{AD}$. Using the same argument as before, we find that quadrilateral $A'B'C'D$ is an isosceles trapezoid and therefore cyclic. So we now have five of the nine points on one circle (i.e., points A', B', C', F, and D).

By repeating the same argument for altitude $\overline{BE}$ (below), we can then state that points D, F, and E lie on the same circle as points A', B', and C'. These six points are as far as Euler got with this configuration.

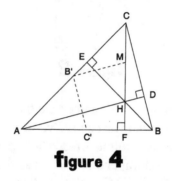

figure 4

With H as the orthocenter (the point of intersection of the altitudes), M is the midpoint of $\overline{CH}$ (see the figure above). Therefore, $\overline{B'M}$, a midline of $\triangle ACH$, is parallel to $\overline{AH}$, or altitude $\overline{AD}$. Since $\overline{B'C'}$ is a midline of $\triangle ABC$, $\overline{B'C'} \parallel \overline{BC}$. Therefore, since $\angle ADC$ is a right triangle, $\angle MB'C'$ is also a right angle. Thus, quadrilateral $MB'C'F$ is cyclic (opposite angles are supplementary). This places point M on the circle determined by points B', C', and F. We now have a seven-point circle.

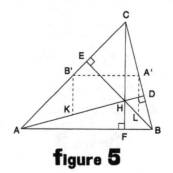

figure 5

We repeat this procedure with point L, the midpoint of $\overline{BH}$ (above). As before, $\angle B'A'L$ is a right angle, as is $\angle B'EL$. Therefore, points B', E, A', and L are concyclic (opposite angles are supplementary). We now have L as an additional point on our circle, making it an eight-point circle.

To locate our final point on the circle, consider point K, the midpoint of $\overline{AH}$. As we did earlier, we find $\angle A'B'K$ to be a right angle, as is $\angle A'DK$. Therefore, quadrilateral $A'DKB'$ is cyclic and point K is on the same circle as points B', A', and D. We have therefore proved that *nine specific points* lie on this circle. This is not to be taken lightly; it is quite spectacular!

5.24. Simson's Invariant

One of the great injustices in the history of mathematics involves a theorem originally published by William Wallace in Thomas Leybourn's *Mathematical Repository* (1799–1800), which through careless misquotes has been attributed to Robert Simson (1687–1768), a famous English interpreter of Euclid's *Elements*. To be consistent with the historic injustice, we shall use the popular reference, and call it Simson's theorem.

The beauty of this theorem lies in its simplicity. Suppose you draw a triangle with its vertices on a circle (something that is always possible, since any three non-collinear points determine a circle) and select a point on the circle that is not at a vertex of the triangle. From that point you draw a perpendicular line to each of the three sides. The three points where these perpendiculars intersect the sides (points X, Y, and Z in the figure below) are always collinear (i.e., they lie on the same straight line). The line that these three points determine is often called the Simson line.

This would be more formally stated in this way:

Simson's theorem: *The feet of the perpendiculars drawn from any point on the circumcircle of a triangle to the sides of the triangle are collinear.*

In the following figure, point P is on the circumcircle of $\triangle ABC$. $\overleftrightarrow{PY} \perp \overleftrightarrow{AC}$ at Y, $\overleftrightarrow{PZ} \perp \overleftrightarrow{AB}$ at Z, and $\overleftrightarrow{PX} \perp \overleftrightarrow{BC}$ at X. According to Simson's (i.e., Wallace's) theorem, points X, Y, and Z are collinear. This line is usually referred to as the Simson line.

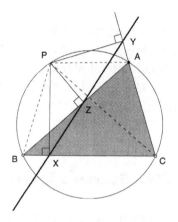

Because of the unconventional nature of the proof of this theorem, it is offered here. Try to follow it. It may be worth the effort. But most important, take time to understand the statement of the theorem.

Proof*: Since $\angle PYA$ is supplementary to $\angle PZA$, quadrilateral $PZAY$ is cyclic.† Draw $\overline{PA}$, $\overline{PB}$, and $\overline{PC}$.

Therefore $\qquad\qquad m\angle PYZ = m\angle PAZ.$ (I)

Similarly, since $\angle PYC$ is supplementary to $\angle PXC$, quadrilateral $PXCY$ is cyclic,

and $\qquad\qquad m\angle PYX = m\angle PCB.$ (II)

However, quadrilateral $PACB$ is also cyclic, since it is inscribed in the given circumcircle, and therefore

---

*For other proofs of Simson's theorem, see Alfred S. Posamentier and Charles T. Salkind, *Challenging Problems in Geometry* (New York: Dover, 1996), pp. 43–45.

†When the opposite angles of a quadrilateral are supplementary (i.e., have a sum of 180°) the quadrilateral is cyclic (i.e., all four points lie on the same circle).

$$m\angle PAZ = m\angle PCB. \qquad (\text{III})$$

From (I), (II), and (III), $m\angle PYZ = m\angle PYX$, and thus points X, Y, and Z are collinear.

This invariant is beautifully demonstrated with Geometer's Sketchpad. There you would draw the figure and then, by moving the point on the circle to various positions, you can observe how the collinearity is preserved under all positions of the point P.

5.25. Ceva's Very helpful relationship

One of the most neglected topics in the high school geometry course is the concept of concurrency. In many cases, it is taken for granted. Oftentimes, we just assume that the altitudes of a triangle are concurrent, that is, they contain a common point of intersection. Similarly, we often just take for granted that the medians of a triangle are concurrent, or the same for the angle bisectors of a triangle. The topic of concurrency of lines in a triangle deserves more attention than it usually gets in an elementary geometry course. In order to put these assumptions to rest, we can establish an extremely useful relationship. This will be done with the help of the famous theorem first published* by the Italian mathematician Giovanni Ceva (1647–1734), and which now bears his name. This relationship is really quite simple and makes a usually difficult topic—concurrency—much less complicated!

In simple terms, the relationship that Ceva established says that if you have three concurrent line segments ($\overline{AL}$, $\overline{BM}$, and $\overline{CN}$, in the following figure), joining a vertex of a triangle with any point on the opposite side, then the products of the alternate segments along the sides (determined by the concurrent segments) are equal, and the converse is true as well. That is, the latter case tells us that if the products of the alternate segments are equal, then the lines from the vertices of the triangle determining those segments are concurrent. In the figure you can see this, noting that the products of the alternate segments along the

---

De lineis se invicem secantibus statica constructio (Milan, 1678).

sides of the triangle are equal: $AN \bullet BL \bullet CM = NB \bullet LC \bullet MA$, since the segments $\overline{AL}$, $\overline{BM}$, and $\overline{CN}$ are concurrent.

Conversely, if $AN \bullet BL \bullet CM = NB \bullet LC \bullet MA$, then the three segments $\overline{AL}$, $\overline{BM}$, and $\overline{CN}$ are concurrent.

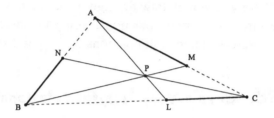

This can be more formally stated as follows:

Ceva's theorem*: *The three lines containing the vertices A, B, and C of $\triangle ABC$ and intersecting the opposite sides at points L, M, and N, respectively, are concurrent if and only if $\frac{AN}{NB} \bullet \frac{BL}{LC} \bullet \frac{CM}{MA} = 1$, or* $AN \bullet BL \bullet CM = NB \bullet LC \bullet MA$.

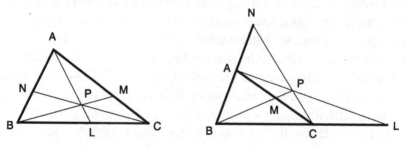

There are two possible situations in which the three lines drawn from the vertices may intersect the sides and still be concurrent. These are pictured in the diagrams above. It is perhaps easier to understand the left-side diagram, and then verify the theorem with the right-side diagram.

Now having accepted this theorem, we can use it. Let's see how simply some of the earlier mentioned relationships can be proved.

---

*The proof of Ceva's theorem is a bit beyond the focus of this book, but it can be found in Alfred S. Posamentier, *Advanced Euclidean Geometry: Excursions for Secondary Teachers and Students* (Emeryville, Calif.: College Publishing, 2002), pp. 27–31.

We shall begin with the task of proving that the medians of a triangle are concurrent. Normally (i.e., without the help of Ceva's theorem) this would be a very difficult proof to do. Because of this, it is often omitted from the high school curriculum. Now observe how simple it is to prove this concurrency.

Proof: In $\triangle ABC$, $\overline{AL}$, $\overline{BM}$, and $\overline{CN}$ are medians. Therefore, $AN = NB$, $BL = LC$, and $CM = MA$. Multiplying these equalities gives us

$$(AN)(BL)(CM) = (NB)(LC)(MA) \text{ or } \frac{AN}{NB} \cdot \frac{BL}{LC} \cdot \frac{CM}{MA} = 1.$$

Thus by Ceva's theorem, $\overline{AL}$, $\overline{BM}$, and $\overline{CN}$ are concurrent.

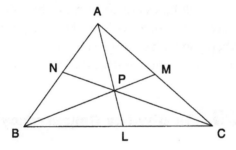

Again, it would be advisable to compare the conventional proof (that which is presented in the context of elementary geometry) for the concurrency of the altitudes of a triangle to the following proof using Ceva's theorem.

We will prove that the altitudes of a triangle are concurrent using Ceva's theorem.

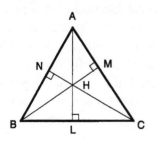

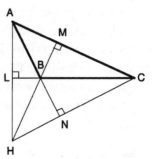

Proof: In $\triangle ABC$, $\overline{AL}$, $\overline{BM}$, and $\overline{CN}$ are altitudes. You may follow this proof for both of the above diagrams, since the same proof holds true for both an acute and an obtuse triangle.

$\triangle ANC \sim \triangle AMB$, so that $\frac{AN}{MA} = \frac{AC}{AB}$ $\qquad$ (I)

$\triangle BLA \sim \triangle BNC$, so that $\frac{BL}{NB} = \frac{AB}{BC}$ $\qquad$ (II)

$\triangle CMB \sim \triangle CLA$, so that $\frac{CM}{LC} = \frac{BC}{AC}$ $\qquad$ (III)

Multiplying (I), (II), and (III) gives us

$$\frac{AN}{MA} \cdot \frac{BL}{NB} \cdot \frac{CM}{LC} = \frac{AC}{AB} \cdot \frac{AB}{BC} \cdot \frac{BC}{AC} = 1.$$

This indicates that the altitudes are concurrent (by Ceva's theorem).

Wouldn't it have been nice to have been exposed to this very useful theorem when you took the high school geometry course? Look how easy it was to prove things previously just accepted.

5.26. An Obvious Concurrency?

A fascinating point of concurrency in a triangle was first established by Joseph-Diaz Gergonne (1771–1859), a French mathematician. Gergonne has a distinct place in the history of mathematics as the initiator (1810) of the first purely mathematical journal, *Annales des mathématiques pures et appliqués*. The journal appeared monthly until 1832 and was known as *Annales del Gergonne*. During the time of its publication, Gergonne published about two hundred papers, mostly on geometry. Gergonne's *Annales* played an important role in the establishment of projective and algebraic geometry since it gave some of the greatest minds of the times an opportunity to share information. We will remember Gergonne for a rather simple theorem that can be stated as follows.

We have a triangle with a circle inscribed in it. The line segments joining the vertices of the triangle with the three points of tangency* are concurrent.

---

*When a circle intersects a line in exactly one point.

To prove this relationship involving concurrency of lines in a triangle we can use Ceva's theorem (from the previous unit).

Gergonne's theorem: *The lines containing a vertex of a triangle and the point of tangency of the opposite side with the inscribed circle are concurrent.* This point of concurrency is known as the *Gergonne Point* of the triangle.

Proof: Circle O is tangent to sides $\overline{AB}$, $\overline{AC}$, and $\overline{BC}$ at points N, M, and L, respectively. It follows that $AN = AM$, $BL = BN$, and $CM = CL$. These equalities may be written as

$$\frac{AN}{AM} = 1, \ \frac{BL}{BN} = 1, \text{ and } \frac{CM}{CL} = 1.$$

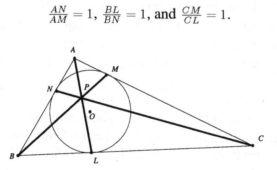

By multiplying these three fractions we get

$$\frac{AN}{AM} \bullet \frac{BL}{BN} \bullet \frac{CM}{CL} = 1$$

Therefore, $\frac{AN}{BN} \bullet \frac{BL}{CL} \bullet \frac{CM}{AM} = 1$, which, as a result of Ceva's theorem (see section 5.25), implies that $\overline{AL}$, $\overline{BM}$, and $\overline{CN}$ are concurrent. This point, P, is the Gergonne Point of $\triangle ABC$.

Neat and (relatively) simple! Yet a fact not well known. These easy-to-understand relationships make geometry fun.

5.27. euler's polyhedra

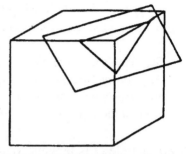

We often see geometric shapes in our daily comings and goings. Yet we don't bother to analyze them for their "hidden" patterns. These can be very. enlightening. Recognizing such patterns will give you a deeper appreciation of the world around you. Leonhard Euler, in the eighteenth century, discovered a lovely relationship among the vertices, faces, and edges of polyhedra.*

You might begin by identifying various polyhedra and counting the number of vertices (V), faces (F), and edges (E), making a chart of these findings and then searching for a pattern.

You may discover that for all these figures, the following relationship holds true: $V + F = E + 2$.

In the cube, the relationship holds true as: $8 + 6 = 12 + 2$.

In the previous figure, if we pass a plane cutting all the edges of a trihedral angle of the polyhedron (i.e., a cube here), we separate one of the vertices from the rest of the polyhedron. But in the process, we add to the polyhedron 1 face, 3 edges, and 3 new vertices. If V is increased by 2, F increased by 1, and E increased by 3, then $V - E + F$ remains unchanged.

That is, $V + F = E + 2 = (8 + 2) + (6 + 1) = (12 + 3) + 2$.

We can obtain a similar result for any polyhedral angle. The new polyhedron will have a new face with the same number of vertices as edges. Since we lose one vertex but gain one face, there is no change in the value of the expression $V - E + F$.

We know the Euler formula applies to a tetrahedron (the "cut off" pyramid has $V + F = E + 2$ and here is $4 + 4 = 6 + 2$). From the above argument, we can conclude that it applies to any polyhedron that can be derived by passing a plane that cuts off a vertex of a tetrahedron a finite number of times. However, we would like it to apply to all simple polyhedra. In the proof we need to show that in regard to the

---

*A polyhedron is a solid figure, bounded by four or more polygonal faces in such a way that pairs meet along edges and three or more edges meet in each vertex. Acube is an example of a polyhedron.

value of the expression $V - E + F$, any polyhedron agrees with the tetrahedron. To do this we need to discuss a new branch of mathematics called topology.*

Topology is a very general type of geometry. The establishment of Euler's formula is a topological problem. Two figures are topologically equivalent if one can be made to coincide with the other by distortion, shrinking, stretching, or bending, but not by cutting or tearing. A teacup and a doughnut are topologically equivalent. The hole in the doughnut becomes the inside of the handle of the teacup.

Topology has been called rubber-sheet geometry. If a face of a polyhedron is removed, the remaining figure is topologically equivalent to a region of a plane. We can deform the figure until it stretches flat on a plane. The resulting figure does not have the same shape or size, but its boundaries are preserved. Edges will become sides of polygonal regions. There will be the same number of edges and vertices in the plane figure as in the polyhedron. Each face of the polyhedron, except the one that was removed, will be a polygonal region in the plane. Each polygon can be cut into triangles, or triangular regions, by drawing diagonals. Each time a diagonal is drawn, we increase the number of edges by 1 but we also increase the number of faces by 1. Hence, the value of $V - E + F$ is undisturbed.

Triangles on the outer edge of the region will have either 1 edge on the boundary of the region, as $\triangle ABC$ in the figure below, or have 2 edges on the boundary, as $\triangle DEF$. We can remove triangles like $\triangle ABC$ by removing the one boundary side. In the figure, this is $\overline{AC}$. This decreases the faces by 1 and the edges by 1. Still, $V - E + F$ is unchanged. If we remove the other kind of boundary triangle, such as $\triangle DEF$, we decrease the number of edges by 2, the number of faces by 1, and the number of vertices by 1. Again, $V - E + F$ *is* unchanged. This process can be continued until one triangle remains.

---

*Topology is the study of geometric forms that remain constant (or invariant) under various transformations, such as stretching and bending.

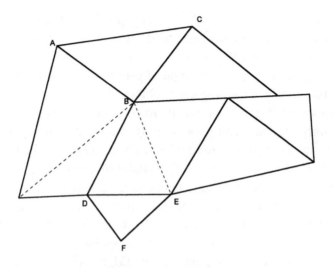

The single triangle has 3 vertices, 3 edges, and 1 face. Therefore, $V - E + F = 1$. Consequently, $V - E + F = 1$ in the plane figure obtained from the polyhedron by distortion. Since one face had been eliminated, we conclude that for the polyhedron

$$V - E + F = 2$$

This procedure applies to any simple polyhedron, even if it is not convex. Can you see why it cannot be applied to a non-simple polyhedron?

An alternative to the approach of distorting the polyhedron to a plane after a face has been eliminated can be named "shrinking a face to a point." If a face is replaced by a point, we lose the n edges of the face and the n vertices of the face, and we lose a face and gain a vertex (the point that replaces the face). This leaves $V - E + F$ unchanged. This process can be continued until only 4 faces remain. Then any polyhedron has the same value for $V - E + F$ as does a tetrahedron. The tetrahedron has 4 faces, 4 vertices, and 6 edges: $4 - 6 + 4 = 2$.

Here is a different type of geometry, and it, too, offers some refreshing delights.

6

ℳathematical 𝒫aradoxes

Paradoxes or fallacies in mathematics generally result from violations of some rules or laws of mathematics. As they violate rules, they also lend themselves to poking some fun at mathematics. Yet in such endeavors we point out the necessity for these rules. One could get the idea (and it may not be entirely wrong) that mathematicians define away inconsistencies so that they can keep the discipline correct.

The paradoxes that are presented in this chapter turn out to be excellent vehicles for presenting these rules of mathematics, for their violation leads to some rather "curious" results, such as 1 = 2, or 1 = 0. Just absurd! They are entertaining since they very subtly lead you to an impossible result. You may become frustrated by the fact that every step seems correct and yet there you are with an absurd result. Again, it is a fine source for investigating the mathematical borders. Why isn't division by zero permissible? Why isn't the product of the radicals always equal to the radical of the product? These are just a few of the questions that this chapter amusingly investigates. The "strange" results are fun to expose.

6.1. are all numbers equal?

The title of this unit is clearly preposterous! But as you will see from the demonstration below, such may not be the case. We shall begin with an easily accepted equation: $\frac{x-1}{x-1} = 1$. Each succeeding row can be justified with elementary algebra. There is nothing wrong with the algebra in the left-hand column. Read this demonstration line by line and then draw your own conclusions.

In terms of x	when $x = 1$
$\frac{x-1}{x-1} = 1$	$\frac{0}{0} = 1$
$\frac{x^2-1}{x-1} = x + 1$	$\frac{0}{0} = 1 + 1 = 2$
$\frac{x^3-1}{x-1} = x^2 + x + 1$	$\frac{0}{0} = 1 + 1 + 1 = 3$
$\frac{x^4-1}{x-1} = x^3 + x^2 + x + 1$	$\frac{0}{0} = 1 + 1 + 1 + 1 = 4$
$\vdots$	
$\frac{x^n-1}{x-1} = x^{n-1} + x^{n-2} + \cdots + x^2 + x + 1$	$\frac{0}{0} = 1 + 1 + 1 + \cdots + 1 = n$

In the above, when $x = 1$, the numbers $1, 2, 3, 4, \ldots, n$ are each equal to $\frac{0}{0}$, which would make them all equal to each other. Of course, this cannot be true. For this reason, we define $\frac{0}{0}$ to be meaningless. To define something to make things meaningful or consistent is what we do in mathematics to avoid ridiculous statements, as was the case here.

6.2. negative One is not equal to Positive One

You need to recall that $\sqrt{2} \bullet \sqrt{3} = \sqrt{6}$, and then perhaps conclude that $\sqrt{a} \bullet \sqrt{b} = \sqrt{ab}$.

From this you should be able to multiply and simplify: $\sqrt{-1} \bullet \sqrt{-1}$.

Some will do the following to simplify this expression: $\sqrt{-1} \bullet \sqrt{-1} = \sqrt{(-1)(-1)} = \sqrt{+1} = 1$.

Others may do the following: $\sqrt{-1} \bullet \sqrt{-1} = (\sqrt{-1})^2 = -1$.

Both "appear" to be correct. But how can this be, that some get -1 and others $+1$?

If both are correct, then $-1 = +1$, since both are equal to $\sqrt{-1} \bullet \sqrt{-1}$. Clearly this can't be true!

What could be wrong? Once again, a fallacy appears when we violate a mathematics rule. Here, we define (for obvious reasons) that $\sqrt{ab} = \sqrt{a} \bullet \sqrt{b}$ is only valid when at least one of a or b is non-negative. This immediately clears up the confusion—the first attempt $\sqrt{-1} \bullet \sqrt{-1} = \sqrt{(-1)(-1)} = \sqrt{+1} = 1$ was wrong since it violates the definition, and the second procedure $\sqrt{-1} \bullet \sqrt{-1} = (\sqrt{-1})^2 = -1$ is right.

6.3. thou Shalt Not divide by Zero

Every mathematician knows that division by zero is forbidden. As a matter of fact, on the list of "commandments" in mathematics, this must certainly be at the top. Why, then, is division by zero not permissible? We in mathematics rely on the order and beauty in which everything in the realm of mathematics falls neatly into place. When something arises that could spoil that order, we simply *define* it to suit our needs. This is precisely what happens with division by zero. One gets a much greater insight into the nature of mathematics by explaining why these "rules" are set forth. So let's give this "commandment" some meaning.

Consider the quotient $\frac{n}{0}$, $n \neq 0$. Without acknowledging the division-by-zero commandment, let us speculate (i.e., guess) what the quotient might be. Let us say it is p, or $\frac{n}{0} = p$. In that case, we could check by multiplying $0 \bullet p$ to see if it equals n, as would have to be the case for the division to be correct. We know that $0 \bullet p \neq n$, since $0 \bullet p = 0$. So there is no number p that can be the quotient for this division. For that reason, we define division by zero to be invalid.

A more convincing case for defining away division by zero is to show how it can lead to a contradiction of an accepted fact, namely, that $1 \neq 2$. We will show that were division by zero acceptable, then 1 would equal 2, clearly an absurdity!

Here is the "proof" that $1 = 2$:

Let $a = b$

Then $a^2 = ab$ [multiplying both sides by a]

$a^2 - b^2 = ab - b^2$ [subtracting b^2 from both sides]

$(a - b)(a + b) = b(a - b)$ [factoring]

$a + b = b$ [dividing by $(a - b)$]

$2b = 1$ [replace a by b, since our premise was $a = b$]

$2 = 1$ [divide both sides by b]

In the step where we divided by $(a - b)$, we actually divided by zero, because $a = b$, so that $a - b = 0$. That ultimately led us to an absurd result, leaving us with no option other than to prohibit division by zero. By taking the time to witness this rule about division by zero you will have a much better appreciation for the nature of mathematics. Then, of course, you will be able to enjoy it because you are slowly being shown not to simply accept everything as "in mathematics it must be just so." A critical eye onto the subject matter is healthy!

6.4. all triangles are isosceles

George Pólya, one of the great mathematicians of our time, said, "Geometry is the science of correct reasoning on incorrect figures." We will observe a demonstration that will draw conclusions based on "incorrect" figures and will lead us to impossible results. Even the statements themselves sound absurd. You may find the demonstration of proving something that is absurd to be either frustrating or enchanting, depending on your disposition. Nevertheless, follow each statement of the "proof" and see if you can detect the mistake. It rests on something that Euclid in his *Elements* would not have been able to resolve because of a lack of a definition.

The fallacy: Any scalene triangle (a triangle with three unequal sides) is isosceles (a triangle having two equal sides).

To prove that scalene △*ABC* is isosceles, we must draw a few auxiliary line segments. Draw the bisector of ∠*C* and the perpendicular bisector of $\overline{AB}$. From their point of intersection, *G*, draw perpendiculars to $\overleftrightarrow{AC}$ and $\overleftrightarrow{CB}$, meeting them at points *D* and *F*, respectively.

It should be noted that there are four possibilities for the above description for various scalene triangles: figure 1, where $\overrightarrow{CG}$ and $\overleftrightarrow{GE}$ meet inside the triangle:

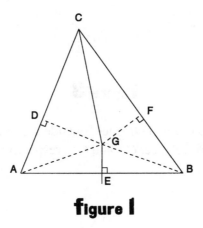

figure 1

figure 2, where $\overrightarrow{CG}$ and $\overleftrightarrow{GE}$ (now actually the midpoint) meet on $\overline{AB}$;

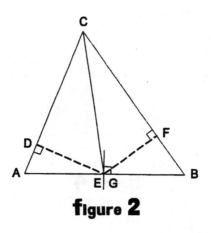

figure 2

figure 3, where $\overrightarrow{CG}$ and $\overleftrightarrow{GE}$ meet outside the triangle, but the perpendiculars $\overline{GD}$ and $\overline{GF}$ fall on $\overline{AC}$ and $\overline{CB}$, respectively;

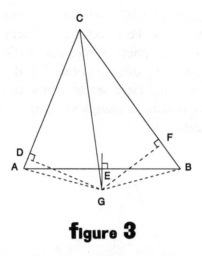

figure 3

and figure 4, where $\overrightarrow{CG}$ and $\overleftrightarrow{GE}$ meet outside the triangle, but the perpendiculars $\overline{GD}$ and $\overline{GF}$ meet $\overrightarrow{CA}$ and $\overrightarrow{CB}$, respectively, outside the triangle.

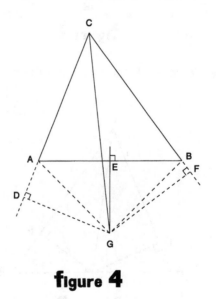

figure 4

Mathematical Paradoxes **219**

The "proof" of the fallacy can be done with any of these figures. Follow the "proof" on any (or all) of these figures.

Given: $\triangle ABC$ is scalene.
Prove: $AC = BC$ (or $\triangle ABC$ is isosceles)

Proof: Since $\angle ACG \cong \angle BCG$ and right $\angle CDG \cong$ right $\angle CFG$, $\triangle CDG \cong \triangle CFG$ (SAA). Therefore, $DG = FG$ and $CD = CF$. Since $AG = BG$ (a point on the perpendicular bisector of a line segment is equidistant from the endpoints of the line segment) and $\angle ADG$ and $\angle BFG$ are right angles, $\triangle DAG \cong \triangle FBG$ (hypotenuse-leg). Thus $DA = FB$.

It then follows that $AC = BC$ (by addition in figures 1 to 3; and by subtraction in figure 4).

At this point you may be somewhat confused, wondering where the error was committed that permitted this fallacy to occur. By rigorous construction, you will find a subtle error in the figures:

a. The point G *must* be outside the triangle.
b. When perpendiculars meet the sides of the triangle, one will meet a side *between* the vertices, while the other will not.

In general terms used by Euclid, this dilemma would remain an enigma, since the concept of *betweenness* was not defined in his *Elements*. In the following discussion we shall prove that errors exist in the fallacious proof above. Our proof uses Euclidean methods, but assumes a definition of betweenness.

Begin by considering the circumcircle of $\triangle ABC$ (see figure 5).

The bisector of $\angle ACB$ must contain the midpoint, G, of $\overset{\frown}{AB}$* (since $\angle ACG$ and $\angle BCG$ are congruent inscribed angles). The perpendicular bisector of $\overline{AB}$ must bisect $\overset{\frown}{AB}$, and therefore pass through G. Thus, the bisector of $\angle ACB$ and the perpendicular bisector of $\overline{AB}$ intersect *outside* the triangle at G. This eliminates the possibilities illustrated in figures 1 and 2.

---

*$\overset{\frown}{AB}$ refers to "arc AB."

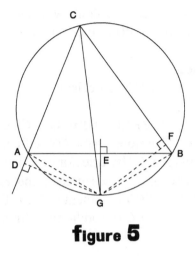

figure 5

Now consider inscribed quadrilateral $ACBG$. Since the opposite angles of an inscribed (or cyclic) quadrilateral are supplementary, $m\angle CAG + m\angle CBG = 180°$. If $\angle CAG$ and $\angle CBG$ are right angles, then $\overline{CG}$ would be a diameter and $\triangle ABC$ would be isosceles. Therefore, since $\triangle ABC$ is scalene, $\angle CAG$ and $\angle CBG$ are not right angles. In this case, one must be acute and the other obtuse. Suppose $\angle CBG$ is acute and $\angle CAG$ is obtuse. Then in $\triangle CBG$ the altitude on $\overline{CB}$ must be *inside* the triangle, while in obtuse $\triangle CAG$, the altitude on $\overrightarrow{CA}$ must be *outside* the triangle. (This is usually readily accepted without proof, but can be easily proved.) The fact that one and only one of the perpendiculars intersects a side of the triangle between the vertices destroys the fallacious "proof."

This rather thorough discussion of this famous geometric fallacy will give you a real appreciation of the precision of geometry. Beyond the entertainment of this unit, it is very instructional, for it gives reason for defining the location of points, something very often neglected.

6.5. an Infinite Series fallacy

Here is a problem that will leave many readers somewhat baffled. Yet the "answer" is a bit subtle and may require some serious thought.

By ignoring the notion of a convergent series* we get the following dilemma:

Let $S = 1 - 1 + 1 - 1 + 1 - 1 + 1 - 1 + \ldots$
$= (1 - 1) + (1 - 1) + (1 - 1) + (1 - 1) + \ldots$
$= 0 + 0 + 0 + 0 + \ldots$
$= 0$

However, were we to group this differently, we would get

Let S $= 1 - 1 + 1 - 1 + 1 - 1 + 1 - 1 + \ldots$
$= 1 - (1 - 1) - (1 - 1) - (1 - 1) - \ldots$
$= 1 - 0 - 0 - 0 - \ldots$
$= 1$

Therefore, since in the first case $S = 1$ and in the second case $S = 0$, we could conclude that $1 = 0$.

What's wrong with this argument?

If this hasn't upset you enough, consider the following argument:

Let $S = 1 + 2 + 4 + 8 + 16 + 32 + 64 + \ldots$ (1)

Here S is clearly positive.

Also, $S - 1 = 2 + 4 + 8 + 16 + 32 + 64 + \ldots$ (2)

Now by multiplying both sides of equation (1) by 2, we get

$2S = 2 + 4 + 8 + 16 + 32 + 64 + \ldots$ (3)

Substituting equation (2) into equation (3) gives us

$2S = S - 1,$

---

*In simple terms, a series converges if it appears to be approaching a specific finite sum. For example, the series $1 + \frac{1}{2} + \frac{1}{4} + \frac{1}{8} + \frac{1}{16} + \frac{1}{32} + \cdots$ converges to 2, while the series $1 + \frac{1}{2} + \frac{1}{3} + \frac{1}{4} + \frac{1}{5} + \frac{1}{6} + \cdots$ does not converge to any finite sum, but continues to grow indefinitely.

from which we can conclude that $S = -1$ (by subtracting S from both sides).

This would have us conclude that -1 is positive, since we established earlier that S was positive.

To clarify the last fallacy, you might want to compare the following correct form of a convergent series:

Let $S = 1 + \frac{1}{2} + \frac{1}{4} + \frac{1}{8} + \frac{1}{16} + \cdots$

We then have $2S = 2 + 1 + \frac{1}{2} + \frac{1}{4} + \frac{1}{8} + \frac{1}{16} + \cdots$

Then $2S = 2 + S$, and $S = 2$, which is true. The difference lies in the notion of a convergent series, as this last one is, while the earlier ones were not convergent and therefore do not allow for the assumptions we made.

6.6. the deceptive border

Have you ever been frustrated by a map book that forced you to turn to the next map page, when the town you were searching for was "just off the map"? Most of these map books, in order to appear attractive, place a border around the map on each page. Have you ever wondered how much space these borders take up?

It probably would be an eye-opener for you to discover this, but, more important, it will make you more alert about the quantitative world around you. Let us consider a map book that has dimensions 8 inches by 10 inches. A modest border might be $\frac{1}{2}$ inch in width and not be considered obtrusive. Let us inspect that situation.

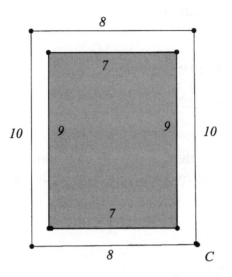

The area of the entire page is 80 square inches and the area of the map is 63 square inches. Therefore, the area of the border region is 80 − 63 = 17 square inches. This happens to be $\frac{17}{80}$ = .2125 = 21.25% or more than $\frac{1}{5}$ the area of the page! Wouldn't it be nice if the "useless" borders did not take up more than 20% of the map book? There would then be fewer pages, and perhaps even a lower cost. Above all, you wouldn't have to turn the page to find your town that just got cut off by the border.

What is essential here is to be alert about the quantitative world around you. There are lots of examples in everyday life that could provoke this kind of astonishment. This sort of eye-opening astonishment should be fun!

6.7. Puzzling Paradoxes

Paradoxes are fun to observe and yet have a very important message imbedded within them. There is much to be learned through this entertainment.

Here are some paradoxes that will give you something to think about.

2 pounds = 32 ounces
$\frac{1}{2}$ pound = 8 ounces

By multiplying the two equalities:

$(2 \bullet \frac{1}{2})$ pound = $(32 \bullet 8)$ ounces
or 1 pound = 256 ounces!

This paradox lies in the fact that the units were not treated properly, and can be best answered by considering the following example:

2 feet = 24 inches
$\frac{1}{2}$ foot = 6 inches

By multiplying we get
1 square foot = 144 square inches.

Another paradox is seen below:
$1 \bullet 0 = 2 \bullet 0$
And we know that $0 = 0$.

Dividing these equalities gives us
$1 = 2$.

Here, of course, we see the familiar rule, not allowing us to divide by 0, being broken and thus leading us to an absurd result.

The messages of each of these paradoxes should remain clear. Don't just dive into an arithmetic process without thinking about what you are doing.

6.8. a trigonometric fallacy

The basis for trigonometry is the Pythagorean theorem. In trigonometry it often manifests itself as $\cos^2 x + \sin^2 x = 1$. You might recall that if a right triangle has sides of lengths sin x and cos x, and hypotenuse equal to 1, then the trigonometric functions hold and the Pythagorean theorem yields $\cos^2 x + \sin^2 x = 1$.

From this, we can show that $4 = 0$. You know this cannot be true! So it is up to you to find the fallacy as it is made. If you don't, we'll expose it at the end of the unit.

The Pythagorean identity can be written as $\cos^2 x = 1 - \sin^2 x$. If we take the square root of each side of this equation, we get

$$\cos x = (1 - \sin^2 x)^{\frac{1}{2}}.$$

We will add 1 to each side of the equation to get

$$1 + \cos x = 1 + (1 - \sin^2 x)^{\frac{1}{2}}.*$$

Now we square both sides: $(1 + \cos x)^2 = [1 + (1 - \sin^2 x)^{\frac{1}{2}}]^2$.

Let us now see what happens when $x = 180°$. $\cos 180° = -1$ and $\sin 180° = 0$.

Substituting into the above equation gives us

$$(1 - 1)^2 = [1 + (1 - 0)^{\frac{1}{2}}]^2$$
$$\text{Then } 0 = (1 + 1)^2 = 4$$

Since $0 \neq 4$, there must be some error. Where is it? Here is a hint:

When $x^2 = p^2$, then $x = +p$ and $x = -p$. The problem situation may call for one or both of these values. Yet sometimes one of them won't work. Look at the step where we took the square root of both sides of the equation. There lies the culprit!

6.9. limits with Understanding

The concept of a limit is not to be taken lightly. It is a very sophisticated concept that can be easily misinterpreted. Sometimes the issues surrounding the concept are quite subtle. Misunderstanding of these can lead to some curious (or humorous, depending on your viewpoint) situations. This can be nicely exhibited with the following two illustrations. Don't be too upset by the conclusion that you will be led to reach. Remember, this is for entertainment. Consider them separately and then notice their connection.

---

*Remember that $x^{\frac{1}{2}} = \sqrt{x}$.

Illustration 1. It is simple to see that the sum of the lengths of the bold segments (the "stairs") is equal to $a + b$.

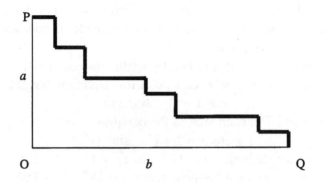

Since the sum of the bold segments ("stairs"), found by summing all the horizontal and all the vertical segments, is $a + b$, when the number of stairs increases, the sum is still $a + b$. The dilemma arises when we increase the stairs to a "limit," so that the set of stairs are so tiny that they appear to be a straight line; in this case the hypotenuse of $\triangle POQ$. It would then appear that $\overline{PQ}$ has length $a + b$. Yet we know from the Pythagorean theorem that $PQ = \sqrt{a^2 + b^2}$ and *not* $a + b$. So what's wrong?

Nothing is wrong! While the set consisting of the stairs as they get smaller does, indeed, approach closer and closer to the straight line segment PQ, it does *not* therefore follow that the *sum* of the bold (horizontal and vertical) lengths approaches the length of $\overline{PQ}$, contrary to intuition. There is no contradiction here, only a failure on the part of our intuition.

Another way to "explain" this dilemma is to argue the following. As the "stairs" get smaller, they increase in number. In an extreme situation, we have 0-length dimensions (for the stairs) used an infinite number of times, which then leads to considering $0 \bullet \infty$, which is meaningless!

A similar situation arises with the following example.

Illustration 2. In the next figure, the smaller semicircles extend from one end of the large semicircle's diameter to the other.

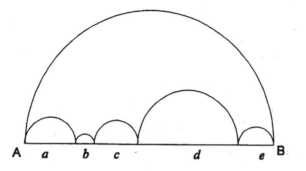

It is easy to show that the sum of the arc lengths of the smaller semicircles is equal to the arc length of the larger semicircle. That is, the sum of the smaller semicircles $= \frac{\pi a}{2} + \frac{\pi b}{2} + \frac{\pi c}{2} + \frac{\pi d}{2} + \frac{\pi e}{2} = \frac{\pi}{2}(a + b + c + d + e) = \frac{\pi}{2}(AB)$, which is the arc length of the larger semicircle. This may not "appear" to be true, but it is! As a matter of fact, as we increase the number of smaller semicircles (where, of course, they get smaller), the sum "appears" to be approaching the length of the segment AB, but, in fact, does not!

Again, the set consisting of the semicircles does indeed approach the length of the straight-line segment AB. It does *not* follow, however, that the *sum* of the semicircles approaches the *length* of the limit, in this case AB.

This "apparent limit sum" is absurd, since the shortest distance between points A and B is the length of segment AB, not the semicircle arc AB (which equals the sum of the smaller semicircles). This is an important concept and may be best explained with the help of these motivating illustrations, so that future misinterpretations can be avoided.

7

Counting and Probability

In today's world, mathematically sophisticated ways of counting are becoming a more important aspect of what we are expected to know in mathematics. More than ever before, concepts of probability are being infused into the things we read: newspapers, magazines, professional reports, and so on. Naturally, these important concepts also have an entertaining side as well. It is that side which we'll savor here.

For example, did you know that the thirteenth of the month is most likely to fall on a Friday, or have you considered the probability of two people in an arbitrarily selected group of thirty people sharing the same birthday? These are just a few of the topics presented in this chapter. It is short and sweet and, hopefully, also entertaining.

7.1. friday the thirteenth!

The number 13 is usually associated with being an unlucky number. Buildings with thirteen or more stories typically will omit the number 13 from the floor numbering. This is immediately noticeable in the elevator, where there is no button for 13. You can certainly think of other examples where the number 13 is associated with bad luck. You likely recall that when the thirteenth of a month turns up on a Friday, then it is particularly bad. This may derive from the belief that there were *thirteen* people present at the Last Supper, which resulted in the crucifixion on a *Friday*.

Do you think that the thirteenth comes up on a Friday with equal regularity as on the other days of the week? You may be astonished that, lo and behold, the thirteenth comes up more frequently on Friday than on any other day of the week.

This fact was first published by B. H. Brown.* He stated that the Gregorian calendar follows a pattern of leap years, repeating every 400 years. The number of days in one 4-year cycle is 3 • 365 + 366. So in 400 years there are 100(3 • 365 + 366) − 3 = 146,097 days. Note that the century year, unless divisible by 400, is not a leap year; hence the deduction of 3. This total number of days is exactly divisible by 7. Since there are 4,800 months in this 400-year cycle, the thirteenth comes up 4,800 times. The following chart summarizes the frequency of the thirteenth appearing on the days of the week.

Day of the week	Number of 13s	Percent
Sunday	687	14.313
Monday	685	14.271
Tuesday	685	14.271
Wednesday	687	14.313
Thursday	684	14.250
Friday	*688*	*14.333*
Saturday	684	14.250

---

*"Solution to Problem E36," *American Mathematical Monthly* 40 (1933): 607.

7.2. think before Counting

Very often a problem situation seems so simple that we plunge right in without first thinking about a strategy to use. This impetuous beginning for the solution often leads to a less elegant solution than one that results from a bit of forethought. Here are two examples of simple problems that can be made even simpler by thinking before working on them.

Find all pairs of prime numbers whose sum equals 999.

Some of you will begin by taking a list of prime numbers and trying various pairs to see if their sum is 999. This is obviously very tedious as well as time consuming, and you would never be quite certain that you had considered all the prime number pairs.

Let's use some logical reasoning to solve this problem. In order to obtain an odd sum for two numbers (prime or otherwise), exactly one of the numbers must be even. Since there is only one even prime, namely 2, there can be only one pair of primes whose sum is 999, and that pair is 2 and 997. That, now, seems so simple.

A second problem where preplanning or some orderly thinking makes sense is as follows:

A palindrome is a number that reads the same forward and backward, such as 747 or 1,991. How many palindromes are there between 1 and 1,000 inclusive?

The traditional approach to this problem would be to attempt to write out all the numbers between 1 and 1,000, and then see which ones are palindromes. However, this is a cumbersome and time-consuming task at best, and one could easily omit some of them.

Let's see if we can look for a pattern to solve the problem in a more direct fashion.

Range	Number of Palindromes	Total Number
1–9	9	9
10–99	9	18
100–199	10	28
200–299	10	38
300–399	10	48
•	•	•
•	•	•
•	•	•

There is a pattern. There are exactly 10 palindromes in each group of 100 numbers (after 99). Thus there will be 9 sets of 10, or 90, plus the 18 from numbers 1 to 99, for a total of 108 palindromes between 1 and 1,000, inclusive.

Another solution to this problem involves organizing the data in a favorable way. Consider all the single-digit numbers (self-palindromes). There are nine such. There are also nine two-digit palindromes. The three-digit palindromes have 9 possible "outside digits" and 10 possible "middle digits,"* so there are 90 of these. In total, there are 108 palindromes between 1 and 1,000, inclusive.

The motto is: think first, then begin a solution!

7.3. the Worthless Increase

Suppose you had a job where you received a 10% raise. Because business was falling off, the boss was soon forced to give you a 10% cut in salary. Will you be back to your starting salary? The answer is a resounding (and very surprising) NO!

This little story is quite disconcerting, since one would expect that with the same percent increase and decrease you should be back to where you started. This is intuitive thinking, but wrong. Convince yourself of this by choosing a specific amount of money and trying to follow the instructions.

---

*Zero can only be a middle digit.

Begin with $100. Calculate a 10% increase on the $\
$110. Now take a 10% decrease of this $110 to get $99—$1\
the beginning amount.

You may wonder whether the result would have been different if we had first calculated the 10% decrease and then the 10% increase. Using the same $100 basis, we first calculate a 10% decrease to get $90. Then the 10% increase yields $99, the same as before. So order makes no difference.

A similar situation, one that is deceptively misleading, can be faced by a gambler. Consider the following situation. You may want to even simulate it with a friend to see if your intuition bears out.

> You are offered a chance to play a game. The rules are simple. There are 100 cards, face down. Of these cards, 55 say *"win"* and 45 say *"lose."* You begin with a bankroll of $10,000. You must bet one-half of your money on each card turned over, and you either win or lose that amount based on what the card says. At the end of the game, all the cards have been turned over. How much money do you have at the end of the game?

The same principle as above applies here. It is obvious that you will win 10 times more than you will lose, so it appears that you will end with more than $10,000. What is obvious is often wrong, and this is a good example. Let's say that you win on the first card; you now have $15,000. Now you lose on the second card; you now have $7,500. If you had first lost and then won, you would still have $7,500. So every time you win one and lose one, you lose one-fourth of your money. So you end up with $10,000 \bullet \left(\frac{3}{4}\right)^{45} \bullet \left(\frac{3}{2}\right)^{10}$.

This is $1.38 when rounded off. Surprised?

7.4. birthday matches

This unit presents one of the most surprising results in mathematics. It is one of the best ways to convince the uninitiated of the power of probability. The results of this unit, aside from being entertaining, will upset your sense of intuition.

Let us suppose that you are in a class with about 35 students. What do you think the chances (or probability) are of two classmates having the same birth date (month and day, only). Intuitively one usually begins to think about the likelihood of 2 people having the same date out of a selection of 365 days (assuming no leap year). Perhaps 2 out of 365? That would be a probability of $\frac{2}{365} = .005479 \approx \frac{1}{2}\%$. A minuscule chance.

Let's consider the "randomly" selected group of the first 35 presidents of the United States. You may be astonished that there are two with the same birth date:

the 11th president, James K. Polk (November 2, 1795), and

the 29th president, Warren G. Harding (November 2, 1865).

You may be surprised to learn that for a group of 35, the probability that two members will have the same birth date is greater than 8 out of 10, or $\frac{8}{10} = 80\%$.

If you have the opportunity, you may wish to try your own experiment by selecting 10 groups of about 35 members to check on date matches. For groups of 30, the probability that there will be a match is greater than 7 out of 10, or in 7 of these 10 groups there ought to be a match of birth dates. What causes this incredible and unanticipated result? Can this be true? It seems to go against our intuition.

To relieve you of your curiosity, we will consider the situation in detail.

Let's consider a class of 35 students. What do you think is the probability that one selected student matches his own birth date? It is a *certainty*, or 1.

This can be written as $\frac{365}{365}$.

The probability that another student does *not* match the first student is $\frac{365-1}{365} = \frac{364}{365}$.

The probability that a third student does *not* match the first or second students is $\frac{365-2}{365} = \frac{363}{365}$.

The probability of all 35 students *not* having the same birth date is the product of these probabilities:

$$p = \frac{365}{365} \cdot \frac{365-1}{365} \cdot \frac{365-2}{365} \cdot \cdots \cdot \frac{365-33}{365} \cdot \frac{365-34}{365}.$$

Since the probability (q) that two students in the group have the same birth date or the probability (p) that two students in the group do *not* have the same birth date is a certainty, the sum of those probabilities must be 1. Thus, $p + q = 1$.

In this case, $q = 1 - \frac{365}{365} \cdot \frac{365-1}{365} \cdot \frac{365-2}{365} \cdot \cdots \cdot \frac{365-33}{365} \cdot \frac{365-34}{365}$ ≈ .8143832388747152. In other words, the probability that there will be a birth date match in a randomly selected group of 35 people is somewhat greater than $\frac{8}{10}$. This is quite unexpected when one considers there were 365 dates from which to choose. The motivated reader may want to investigate the nature of the probability function. Here are a few values to serve as a guide:

Number of people in group	Probability of a birth date match
10	.1169481777110776
15	.2529013197636863
20	.4114383835805799
25	.5686997039694639
30	.7063162427192686
35	.8143832388747152
40	.8912318098179490
45	.9409758994657749
50	.9703735795779884
55	.9862622888164461
60	.9941226608653480
65	.9976831073124921
70	.9991595759651571

Notice how quickly "almost-certainty" is reached. With about 60 students in a room the chart indicates that it is almost certain (greater than 99%) that two students will have the same birth date.

Were one to do this with the death dates of the first 35 presidents, one would notice that two died on March 8 (Millard Fillmore in 1874 and William H. Taft in 1930) and three died on July 4 (John Adams and Thomas Jefferson in 1826, and James Monroe in 1831).

Above all, this astonishing demonstration should serve as an eye-opener about the inadvisability of relying too much on intuition.

7.5. Calendar Peculiarities

The calendar holds many recreational ideas that can be exploited to turn students on to mathematics—at least to explore number relationships.
    Consider any calendar page, say, October 2002.

Sunday	Monday	Tuesday	Wednesday	Thursday	Friday	Saturday
		1	2	3	4	5
6	7	8	9	10	11	12
13	14	15	16	17	18	19
20	21	22	23	24	25	26
27	28	29	30	31		

Select a (3×3) square of any nine dates on the calendar. We will select those shaded above. Add 8 to the smallest number in the shaded region and then multiply by 9. So $(9 + 8) \cdot 9 = 153$. Then multiply the sum of the numbers of the middle row (51) of this shaded matrix by 3. Surprise! It is the same as the previous answer, 153. But why? Here are some clues: the middle number is the mean (or average) of the 9 shaded numbers. The sum of the numbers in the middle column is $\frac{1}{3}$ of the sum of the nine numbers. This investigation will have favorable results.

    Now that you have an appreciation for the calendar, consider what the probability is of 4/4*, 6/6, 8/8, 10/10, and 12/12 all falling on the same day of the week. More than likely the knee-jerk reaction will be about $\frac{1}{5}$. Wrong! The probability is 1, a certainty! But why this surprising result? Closer inspection will reveal that these dates are all exactly nine weeks apart. Such little known facts always draw an interest that otherwise would be untapped. Counting on a calendar presents numerous surprises.

---

*4/4 represents April 4, 6/6 represents June 6, and so on.

7.6. the monty hall problem
("Let's Make a Deal")

Let's Make a Deal was a long-running television game show that featured a problematic situation. A randomly selected audience member would come on stage and be presented with three doors. She was asked to select one, hopefully the one behind which there was a car and not one of the other two doors, each of which had a donkey behind it. There was only one wrinkle in this: after the contestant made her selection, the host, Monty Hall, exposed one of the two donkeys behind a not-selected door (leaving two doors still unopened) and the audience participant was asked if she wanted to stay with her original selection (not yet revealed) or switch to the other unopened door. At this point, to heighten the suspense, the rest of the audience would shout out "stay" or "switch" with seemingly equal frequency. The question is what to do? Does it make a difference? If so, which is the better strategy (i.e., the greater probability of winning) to use here?

Let us look at this now step-by-step. The result gradually will become clear.

There are *two donkeys* and *one car* behind these doors
You must try to get the car. You select door #3

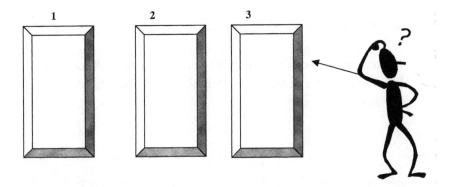

Monty Hall opens one of the doors that you *did not* select and exposes a donkey.

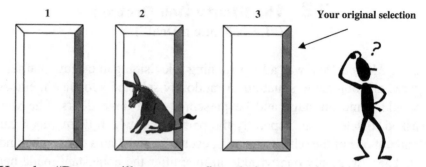

He asks: "Do you still want your first choice door, or do you want to switch to the other closed door?"

To help make a decision, consider an *extreme case*:

Suppose there were 1,000 doors instead of just three doors.

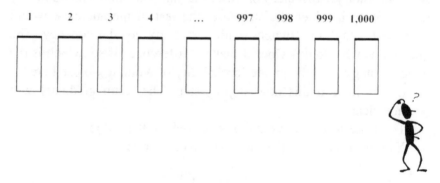

You choose door # 1000. How likely is it that you chose the right door?

Very unlikely, since the probability of getting the right door is $\frac{1}{1,000}$

How likely is it that the car is behind one of the other doors?

Very likely: $\frac{999}{1,000}$

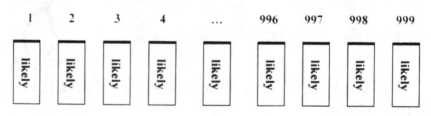

These are all *very likely* doors!

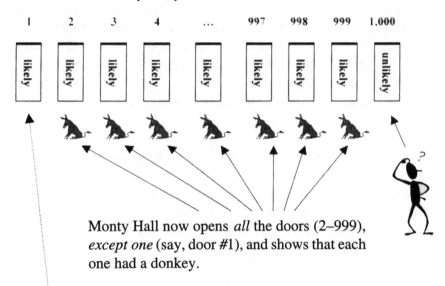

Monty Hall now opens *all* the doors (2–999),
except one (say, door #1), and shows that each
one had a donkey.

A "very likely" door is left: door #1.
We are now ready to answer the question. Which is a better choice:

♦ door # 1,000 ("Very unlikely" door), or
♦ door # 1 ("Very likely" door)?

The answer is now obvious. We ought to select the "very likely" door, which means "switching" is the better strategy for the audience participant to follow.

In the extreme case, it is much easier to see the best strategy, than had we tried to analyze the situation with the three doors. The principle is the same in either situation.

This problem has caused many an argument in academic circles and was also a topic of discussion in the *New York Times*, and other popular publications as well. John Tierney wrote in the *New York Times* (Sunday, July 21, 1991) that "perhaps it was only an illusion, but for a moment here it seemed that an end might be in sight to the debate raging among mathematicians, readers of *Parade* magazine, and fans of the television game show *Let's Make a Deal*. They began arguing last September after Marilyn vos Savant published a puzzle in *Parade*. As readers of her 'Ask Marilyn' column are reminded each week, Ms. vos Savant is listed in the Guinness Book of World Records Hall of Fame for 'Highest IQ,' but that credential did not impress the public when she answered this question from a reader." She gave the right answer, but still many mathematicians argued.

7.7. anticipating heads and tails

This lovely little unit will show you how some clever reasoning, along with algebraic knowledge of the most elementary kind, will help you solve a seemingly "impossibly difficult" problem.

Consider the following problem:

> You are seated at a table in a dark room. On the table, there are 12 pennies, 5 of which are heads up and 7 are tails up. (You know where the coins are, so you can move or flip any coin, but because it is dark you will not know if the coin you are touching was originally heads up or tails up.) You are to separate the coins into two piles (possibly flipping some of them) so that when the lights are turned on there will be an equal number of heads in each pile.

Your first reaction is "you must be kidding!" "How can anyone do this task without seeing which coins are heads or tails up?" This is where a most clever (yet incredibly simple) use of algebra will be the key to the solution.

Let's "cut to the chase." Here is what you do. (You might actually want to try it with 12 coins.) Separate the coins into two piles, of 5 and 7 coins, respectively. Then flip over the coins in the smaller pile. Now both piles will have the same number of heads! That's all! You will think this is magic. How did this happen? Well, this is where algebra helps us understand what was actually done.

Let's say that when you separate the coins in the dark room, h heads will end up in the 7-coin pile. Then the other pile, the 5-coin pile, will have $5 - h$ heads and $5 - (5 - h) = h$ tails. When you flip all

the coins in the smaller pile, the $5 - h$ heads (because ⌐
heads up altogether) become tails and the h tails become h hea⌐
each pile contains h heads!

8

Mathematical Potpourri

a ll the topics that could not find a proper home in the first seven chapters of this book reside in this chapter. We have a mixture of delightful topics that are sure to turn you on to mathematics. Do not be fooled by their location; less important topics are not relegated to the last chapter. Quite the contrary.

Here you will see one of the most amazing magic squares, whose first appearance was in Albrecht Dürer's *Melencolia I*. It has a plethora of properties above and beyond those of normal magic squares. You will be exposed to mathematical manifestations in nature, and you will be presented ultimately with some famous unsolved problems (no, you are not expected to solve problems that, for hundreds of years, have not been solved). It is quite likely that this last chapter could prove most entertaining, since it seems to cover a very wide range of topics, none of which allows itself to be categorized in the previous seven chapters. Perhaps we should have called this chapter one!

8.1. perfection In mathematics

What is perfect in mathematics, a subject where most think everything is already perfect? Over the years, various authors have been found to determine perfect squares, perfect numbers, perfect rectangles, and perfect triangles. You might try to add to the list of "perfection." What other mathematical things may be worthy of the adjective "perfect"?

Begin with the *perfect squares*. They are well known: 1, 4, 9, 16, 25, 36, 49, 64, 81, 100, . . . They are numbers whose square roots are natural numbers: 1, 2, 3, 4, 5, 6, 7, 8, 9, 10, . . .

A *perfect number* is one that equals the sum of its factors (excluding the number itself). The first four perfect numbers are

6 [1 + 2 + 3],
28 [1 + 2 + 4 + 7 + 14],
496 [1 + 2 + 4 + 8 + 16 + 31 + 62 + 124 + 248], and
8,128 [This is left to the reader to verify].

Perfect numbers were already known to the ancient Greeks (*Introductio Arithmeticae* by Nichomachus, ca. 100 C.E.). Interestingly, the Greeks felt that there was exactly one perfect number for each digit-group of numbers. The first four perfect numbers seemed to fit this pattern, namely, among the single-digit numbers the only perfect number is 6, among the two-digit numbers there was only 28, then 496 was the only three-digit perfect number, and 8,128 was the only four-digit perfect number. What would you predict is the number of digits in the next larger perfect number? You will probably say it must be a five-digit number. Furthermore, if you were to make other conjectures about perfect numbers, you may conclude that perfect numbers must end in a 6 or an 8, alternately.

As a matter of fact, there is no five-digit perfect number at all. This should teach you to be cautious about making predictions with relatively little evidence. The next larger perfect number has 8 digits: 33,550,336. Then we must take a large leap to the next perfect number: 8,589,869,056. Here we also see that our conjecture (although reasonable) of getting alternate final digits of 6 and 8 is false.* This is a good lesson about drawing inductive conclusions prematurely.

A *perfect rectangle* is one that has an area numerically equal to its perimeter. There are only two perfect rectangles, namely, one having sides of lengths 3 and 6, and the other with all sides of length 4.

There are also *perfect triangles.*† These are defined as triangles whose areas are numerically equal to their perimeters. You should be able to identify the right triangles that fit that pattern by simply setting the area and perimeter formulas equal to each other. Among the right triangles there are only two perfect triangles, one with sides of lengths 6, 8, 10 and the other with sides of lengths 5, 12, 13.

Among the non-right triangles, there are only three whose areas are numerically equal to their perimeters. Their sides measure 6, 25, 29; 7, 15, 20; and 9, 10, 17.

The three above cases can be verified using Heron's formula.‡

What does this do for us? Very little, except to allow us to appreciate the "perfection" in mathematics. Perhaps you will be encouraged to find other candidates in mathematics deserving the label *perfection*.

8.2. the beautiful magic Square

There are entire books written about magic squares§ of all kinds. There is one magic square, however, that stands out from the rest for

---

* The formula for a perfect number is the following.

If $2^k - 1$ is a prime number ($k > 1$), then $2^{k-1} (2^k - 1)$ is an even perfect number.

†See M. V. Bonsangue et al., "In Search of Perfect Triangles," *Mathematics Teacher* 92, no.1 (January 1999): 56–61.

‡The formula for the area of a triangle, given only the lengths of its sides, was accredited to Heron of Alexandria (150 B.C.E.) and reads Area = $\sqrt{s(s - a)(s - b)(s - c)}$, where a, b, and c are the lengths of the sides and s is the semiperimeter.

§Two recommended books are: W. H. Benson, and O. Jacoby, *New Recreations with Magic Squares* (New York: Dover, 1976) and W. S. Andrews, *Magic Squares and Cubes* (New York: Dover, 1960). A concise treatment can be found in A. S. Posamentier and J. Stepelman, *Teaching Secondary School Mathematics: Techniques and Enrichment Units*, 6th ed. (Upper Saddle River, N.J.: Prentice Hall/Merrill, 2002), pp. 240–44.

its beauty. This one special square has many properties beyond that required for a square matrix of numbers to be considered "magic." This magic square even comes to us through art, and not through the usual mathematical channels. It is depicted in the background of the famous engraving produced in 1514 by the renowned German artist Albrecht Dürer (1471–1528), who lived in Nürnberg, Germany.

A magic square is a square matrix of numbers, where the sum of the numbers in each of its columns, rows, and diagonals is the same. Most of Dürer's works were signed by him with his initials, one over the other, plus the year in which the work was made. Here we find it near the lower right side of the picture. We notice that it was made in the year 1514. The observant reader may notice that the two center cells of the bottom row of the magic square depict the year as well. Let us look at this magic square more closely.

16	3	2	13
5	10	11	8
9	6	7	12
4	15	14	1

First, let's make sure that it is a magic square.* The sum of all the rows, all the columns, and diagonals must be equal. Well, they have a sum of 34. So that is all that would be required for this square matrix of numbers to be considered a "magic square." However, this Dürer magic square has lots more properties that other magic squares do not have. We shall list some here:

- The four corner numbers have a sum of 34.
    16 + 13 + 1 + 4 = 34

- Each of the corner 2 by 2 squares has a sum of 34.
    16 + 3 + 5 + 10 = 34
    2 + 13 + 11 + 8 = 34
    9 + 6 + 4 + 15 = 34
    7 + 12 + 14 + 1 = 34

---

*A 4 by 4 magic square is usually constructed by writing the numbers from 1 to 16 in proper order, row by row, and then striking out the numbers in the two diagonals. Each of these struck out numbers is then to be replaced by its complement, that is, the number which when added to it yields a sum of 17 (one greater than the number of cells). However, the Dürer square interchanged the two middle columns to get the date of the etching in the two bottom center cells.

- The center 2 by 2 square has a sum of 34.

  $10 + 11 + 6 + 7 = 34$

- The sum of the numbers in the diagonal cells equals the sum of the numbers in the cells not in the diagonals.

  $16 + 10 + 7 + 1 + 4 + 6 + 11 + 13 = 3 + 2 + 8 + 12 + 14 + 15 + 9 + 5 = 68$

- The sum of the squares of the numbers in the diagonal cells equals the sum of the squares of the numbers not in the diagonal cells.

  $16^2 + 10^2 + 7^2 + 1^2 + 4^2 + 6^2 + 11^2 + 13^2 = 3^2 + 2^2 + 8^2 + 12^2 + 14^2 + 15^2 + 9^2 + 5^2 = 748$

- The sum of the cubes of the numbers in the diagonal cells equals the sum of the cubes of the numbers not in the diagonal cells.

  $16^3 + 10^3 + 7^3 + 1^3 + 4^3 + 6^3 + 11^3 + 13^3 = 3^3 + 2^3 + 8^3 + 12^3 + 14^3 + 15^3 + 9^3 + 5^3 = 9,248$

- The sum of the squares of the numbers in the diagonal cells equals the sum of the squares of the numbers in the first and third rows.

  $16^2 + 10^2 + 7^2 + 1^2 + 4^2 + 6^2 + 11^2 + 13^2 = 16^2 + 3^2 + 2^2 + 13^2 + 9^2 + 6^2 + 7^2 + 12^2 = 748$

- The sum of the squares of the numbers in the diagonal cells equals the sum of the squares of the numbers in the second and fourth rows.

  $16^2 + 10^2 + 7^2 + 1^2 + 4^2 + 6^2 + 11^2 + 13^2 = 5^2 + 10^2 + 11^2 + 8^2 + 4^2 + 15^2 + 14^2 + 1^2 = 748$

- The sum of the squares of the numbers in the diagonal cells equals the sum of the squares of the numbers in the first and third columns.

  $16^2 + 10^2 + 7^2 + 1^2 + 4^2 + 6^2 + 11^2 + 13^2 = 16^2 + 5^2 + 9^2 + 4^2 + 2^2 + 11^2 + 7^2 + 14^2 = 748$

- The sum of the squares of the numbers in the diagonal cells equals the sum of the squares of the numbers in the second and fourth columns.

$16^2 + 10^2 + 7^2 + 1^2 + 4^2 + 6^2 + 11^2 + 13^2 = 3^2 + 10^2 + 6^2 + 15^2 + 13^2 + 8^2 + 12^2 + 1^2 = 748$

- Notice the following beautiful symmetries:

$2 + 8 + 9 + 15 = 3 + 5 + 12 + 14 = 34$

$2^2 + 8^2 + 9^2 + 15^2 = 3^2 + 5^2 + 12^2 + 14^2 = 374$

$2^3 + 8^3 + 9^3 + 15^3 = 3^3 + 5^3 + 12^3 + 14^3 = 4,624$

- The sum of each adjacent left and right pair of numbers produces a pleasing symmetry:

$16 + 5 = \mathbf{21}$ $3 + 10 = \mathbf{13}$ $2 + 11 = \mathbf{13}$ $13 + 8 = \mathbf{21}$

$9 + 4 = \mathbf{13}$ $6 + 15 = \mathbf{21}$ $7 + 14 = \mathbf{21}$ $12 + 1 = \mathbf{13}$

- The sum of each adjacent left and right pair of numbers produces a pleasing symmetry:

$16 + 3 = \mathbf{19}$ $2 + 13 = \mathbf{15}$

$5 + 10 = \mathbf{15}$ $11 + 8 = \mathbf{19}$

$9 + 6 = \mathbf{15}$ $7 + 12 = \mathbf{19}$

$4 + 15 = \mathbf{19}$ $14 + 1 = \mathbf{15}$

Can you find some other patterns in this beautiful magic square? Remember this is not a typical magic square. There all that is required is that all the rows, columns, and diagonals have the same sum. In the Dürer magic square there are many more properties.

8.3. Unsolved Problems

Who says that all mathematical problems get solved? Unsolved problems have a very important role in mathematics. Attempts to solve them oftentimes lead to very important findings of other sorts. An unsolved problem—one not yet solved by the world's most brilliant minds—tends to pique our interest by quietly asking us if we can solve it, especially when the problem itself is exceedingly easy to understand. We shall look at some unsolved problems to get a better understanding of the history of mathematics. Twice in recent years, mathematics has made newspaper headlines, each time with the solution to a longtime unsolved problem.

The *Four-Color Map Problem* dates back to 1852, when Francis Guthrie, while trying to color the map of the counties of England, noticed that four colors sufficed. He asked his brother Frederick if it was true that *any* map can be colored using four colors in such a way that adjacent regions (i.e., those sharing a common boundary line, not just a point) receive different colors. Frederick Guthrie then communicated the conjecture to the famous mathematician Augustus DeMorgan. In 1977 the "four-color map" problem was solved by two mathematicians, K. Appel and W. Haken, who, using a computer, considered all possible maps and established that it was never necessary to use more than four colors to color a map so that no two territories sharing a common border would be represented by the same color.

More recently, on June 23, 1993, Andrew Wiles, a Princeton University mathematics professor, announced that he solved the 350-year-old "Fermat's last theorem." It took him another year to fix some errors in the proof, but it puts to rest a nagging problem that occupied scores of mathematicians for centuries. The problem, which Pierre de Fermat wrote (ca. 1630) in the margin of a mathematics book (Diophantus' *Arithmetica*) he was reading, was discovered by his son after his death. In addition to the statement of the theorem, Fermat stated that his proof was too long to fit the margin, so he effectively left to others the job of proving his statement. To this day we speculate if Fermat really did have a proof, or if this was put there as a joke for posterity. Some say that he may have thought he had a proof, but it might not have been correct. In any case, let's take a look at what this famous theorem stated.

Fermat's theorem: $x^n + y^n = z^n$ *has no non-zero integer solutions for* $n > 2$.*

---

*In other words, there are no integer values for x, y, and z for which this equation will hold true, with the exception when integer $n \leq 2$.

During this time, speculation began about other unsolved problems, of which there are still many. Two of them are very easy to understand, but apparently exceedingly difficult to prove. Neither has yet been proved.

Christian Goldbach (1690–1764), a Prussian mathematician, in a 1742 letter to the famous Swiss mathematician Leonhard Euler, posed the following problem, which to this day has yet to be solved.

Goldbach's conjecture: *Every even number greater than 2 can be expressed as the sum of two prime numbers.*

Here is a list of some even numbers and their corresponding sum of two prime numbers.

Even numbers greater than 2	Sum of two prime numbers
4	2 + 2
6	3 + 3
8	3 + 5
10	3 + 7
12	5 + 7
14	7 + 7
16	5 + 11
18	7 + 11
20	7 + 13
48	19 + 29
100	3 + 97

Can you find some more examples of this?

Goldbach's second conjecture: *Every odd number greater than 5 can be expressed as the sum of three primes.*

Let us consider the first few odd numbers:

Odd numbers greater than 5	Sum of three prime numbers
7	2 + 2 + 3
9	3 + 3 + 3
11	3 + 3 + 5
13	3 + 5 + 5
15	5 + 5 + 5
17	5 + 5 + 7
19	5 + 7 + 7
21	7 + 7 + 7
51	3 + 17 + 31
77	5 + 5 + 67
101	5 + 7 + 89

You may wish to see if there is a pattern here and generate other examples.

These unsolved problems have tantalized many mathematicians over the centuries and although no solutions have yet been found, there is more evidence (with the help of computers) that these must be true since no counterexamples have been found. Interestingly, the efforts to solve these have led to some important discoveries in mathematics that might have remained hidden without this impetus. For us they provide sources for entertainment.

8.4. an Unexpected result

Here is one with which you can have fun. Consider the following sequence and find the next number: **1, 2, 4, 8, 16**. Clearly most people would guess that 32 is the next number. Yes, that would be fine. However, when the next number is given as 31 (instead of the expected 32), cries of "wrong!" are usually heard.

Much to your amazement, this is a correct answer, and **1, 2, 4, 8, 16, 31** can be a legitimate sequence.

The task now is to convince you of the legitimacy of this sequence. It would be nice if it could be done geometrically, as that would give convincing evidence of a physical nature. We will do that later, but in the meantime let us first find the succeeding numbers in this "curious" sequence.

We shall set up a table of differences (i.e., a chart showing the differences between terms of a sequence), beginning with the given sequence up to 31, and then work backward once a pattern is established (here at the third difference).

Original Sequence	1	2	4	8	16	31
First Difference		1	2	4	8	15
Second Difference			1	2	4	7
Third Difference				1	2	3
Fourth Difference					**1**	**1**

With the fourth differences forming a sequence of constants, we can reverse the process (turn the table upside down), and extend the third differences a few more steps with 4 and 5.

Fourth Difference				1	1	1	1	
Third Difference			1	2	3	**4**	**5**	
Second Difference		1	2	4	7	**11**	**16**	
First Difference		1	2	4	8	**15**	**26**	**42**
Original Sequence	1	2	4	8	16	**31**	**57**	**99**

The boldface numbers are those that were obtained by working backward from the third difference sequence. Thus, the next numbers of the given sequence are: 57 and 99. The general term is a fourth-power expression since we had to go to the third differences to get a constant. The general term (n) is $\frac{n^4 - 6n^3 + 23n^2 - 18n + 24}{24}$

One should not think that this sequence is independent of other parts of mathematics.

Consider the Pascal triangle:*

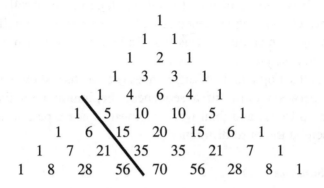

```
                    1
                 1     1
              1     2     1
           1     3     3     1
          1     4     6     4     1
       1 \  5    10    10     5     1
     1     6 \ 15    20    15     6     1
   1     7    21 \ 35    35    21     7     1
 1     8    28    56 \ 70    56    28     8     1
```

Consider the horizontal sums of the rows of the Pascal triangle to the right of the bold line drawn: 1, 2, 4, 8, 16, 31, 57, 99, 163. This is again our newly developed sequence.

A geometric interpretation should help convince you of the beauty and consistency inherent in mathematics. To do this we shall make a chart of the number of regions into which a circle can be partitioned by joining points on the circle. You ought to try this by partitioning a circle. Just make sure that no three lines meet at one point, or else you will lose a region.

Number of points on the circle	Number of regions into which the circle is partitioned
1	1
2	2
3	4
4	8
5	16
6	31
7	57
8	99

---

*This triangle is formed by beginning in the first row with 1, then the second row has 1, 1. The third row is obtained by placing 1s at the ends and adding the two numbers in the second row (1 + 1 = 2) to get the 2. The fourth row is obtained the same way. After the end 1s are placed, the 3s are derived from the sum of the two numbers above (to the right and left), that is, 1 + 2 = 3 and 2 + 1 = 3.

Now that you see that this unusual sequence appears in various other fields, you may feel a degree of satisfaction emerging.

8.5. mathematics in nature

We are now about to embark on one of the most beautiful phenomena in mathematics. It truly belongs in this chapter since it permeates more fields of mathematics than almost any other topic. The famous Fibonacci numbers, a sequence of numbers that was the direct result of a problem posed by Leonardo of Pisa (known commonly as Fibonacci),* in his book *Liber abaci* (1202), regarding the regeneration of rabbits has many other applications in nature. At first sight, it may appear that these applications are coincidental, but eventually you will be amazed at the vastness of the appearance of this famous sequence of numbers.

The original problem posed by Fibonacci asks for the number of pairs of rabbits accumulating each month and leads to the sequence:

$$1, 1, 2, 3, 5, 8, 13, 21, 34, 55, 89, 144, \ldots$$

To best enchant you with the many applications of the Fibonacci numbers, you ought to have various species of pinecones, a pineapple, a plant (see below) and, if possible, other spiral examples in nature (e.g., a sunflower). The number of spirals on each of these will be a Fibonacci number.

Can you determine a relationship between the Fibonacci numbers and the leaves of a plant (have a plant on hand)? From the standpoint of Fibonacci numbers, one may observe two items: (1) the number of leaves it takes to go (rotating about the stem) from any given leaf to the next one "similarly placed" (i.e., above it and in the same direction) on the stem; and (2) the number of revolutions as one follows the leaves in going from one leaf to another one "similarly placed." In both cases, these numbers turn out to be the Fibonacci numbers.

In the case of leaf arrangment, the following notation is used: $\frac{3}{8}$

---

*See section 1.18.

means that it takes three revolutions and eight leaves to arrive at the next leaf "similarly placed." In general, if we let *r* equal the number of revolutions, and *s* equal the number of leaves it takes to go from any given leaf to one "similarly placed," then $\frac{r}{s}$ will be the *phyllotaxis* (the arrangement of leaves in plants). Look at the figure below and try to find the plant ratio.

In this figure, the plant ratio is $\frac{5}{8}$:

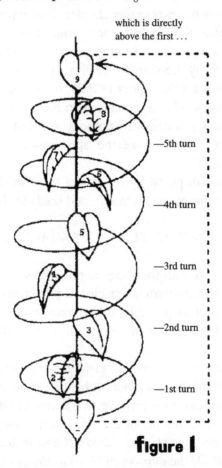

which is directly
above the first ...

—5th turn

—4th turn

—3rd turn

—2nd turn

—1st turn

figure I

The pinecone also presents a Fibonacci application. The bracts on the cone are considered to be modified leaves, compressed into a smaller space. Upon observation of the cone, one can notice two spirals, one to the left (clockwise) and the other to the right (counterclockwise).

One spiral increases at a sharp angle, while the other spiral increases more gradually. Consider the steep spirals and count them as well as the spirals that increase gradually. Both numbers should be Fibonacci numbers. For example, a white pinecone has five clockwise spirals and eight counterclockwise spirals. Other pinecones may have different Fibonacci ratios. If you have a chance, examine the daisy or sunflower to see where the Fibonacci ratios apply to them.

We notice that the ratios of consecutive Fibonacci numbers approach the Golden Ratio (see section 5.12).

If we look closely at the ratios of consecutive Fibonacci numbers, we can approximate their decimal equivalents. The early Fibonacci ratios are

$$\frac{2}{3} = .666667$$
$$\frac{3}{5} = .600000$$

Then, as we go further along the sequence of Fibonacci numbers, the ratios begin to approach ϕ:

$$\frac{89}{144} = .618056$$
$$\frac{144}{233} = .618026$$

The Golden Ratio, $\phi = .6180339887498948482045868343 6564\ldots$

Geometrically, point B in figure 2 divides $\overline{AC}$ into the Golden Ratio, $\frac{AB}{BC} = \frac{BC}{AC} \approx .618034$

A   B   C

figure 2

Now consider the series of Golden Rectangles, those whose dimensions are chosen so that the ratio of $\frac{width}{length}$ is the Golden Ratio $\frac{w}{l} = \frac{l}{w+l}$.

math *Charmers*

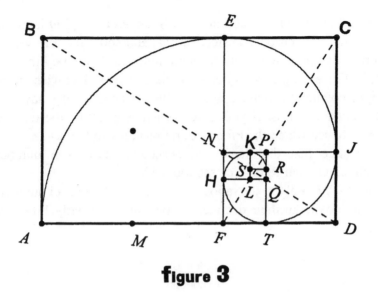

figure 3

If the rectangle is divided by a line segment ($\overline{EF}$) into a square ($ABEF$) and a Golden Rectangle ($EFDC$), and if we keep partitioning each new Golden Rectangle in the same way, we can construct a "logarithmic spiral" by drawing quarter-circles as shown above in the successive squares as shown above. This type of curve is frequently found in the arrangements of seeds in flowers or in the shapes of seashells and snails. If possible, you ought to find illustrations to show these spirals.

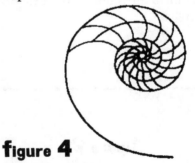

figure 4

For another example of mathematics in nature, you should consider the pineapple. Here there are three distinct spirals of hexagons: a group of *5* spirals winding gradually in one direction, a second group of *13* spirals winding more steeply in the same direction, and a third group of *8* spirals winding in the opposite direction. The number of spirals in each group is a Fibonacci number. Figure 5 shows a representation of the pineapple with the scales numbered in order. This order is determined by the relative distance each hexagon is from the bottom. That is, the lowest is numbered 0, the next higher one is numbered 1. Note hexagon 42 is slightly higher than hexagon 37.

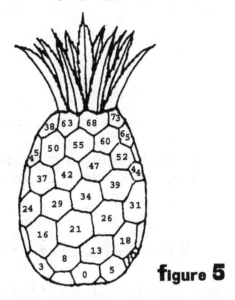

figure 5

Note the three distinct sets of spirals in figure 5 that cross each other, starting at the bottom. One spiral is the 0, 5, 10, etc., sequence, which increases at a slight angle. The second spiral is the 0, 13, 26, etc., sequence, which increases at a steeper angle. The third spiral has the 0, 8, 16, etc., sequence, which lies in the opposite direction from the other two. Try to figure out the common difference between the numbers in each sequence. In this case, the differences are 5, 8, and 13, all of which are Fibonacci numbers. Different pineapples may have different sequences.

Not to be cute, but to move these applications to a completely different venue, consider the regeneration of male bees. We shall accept that male bees hatch from unfertilized eggs, female bees from fertilized eggs. The flow chart below will help you trace the regeneration of the male bees. The following pattern develops:

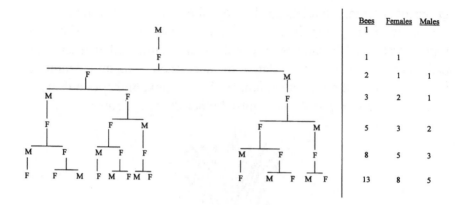

Bees	Females	Males
1		
1	1	
2	1	1
3	2	1
5	3	2
8	5	3
13	8	5

It should be obvious by now that this pattern is the Fibonacci sequence.

As was said earlier, there are practically endless applications of the Fibonacci numbers. This just adds to the amazement these applications usually generate.

8.6. the hands of a Clock

The clock can be an interesting source of mathematical applications. These applications are focused on mathematics, rather than the more common use of applications of mathematics in other disciplines. The clock is a good source for some entertainment with mathematics.

Begin by determining the *exact time* that the hands of a clock will overlap after 4:00. Your first reaction to the solution to this problem is likely to be that the answer is simply 4:20.

That does not take into account that the hands move uniformly while the minute hand moves faster. With this in mind, the astute reader will begin to estimate the answer to be between 4:21 and 4:22. You should realize that the hour hand moves through an interval between minute markers every 12 minutes. Therefore it will leave the interval 4:21–4:22 at 4:24. This, however, doesn't answer the original question about the exact time of this overlap.

Let's consider a technique to better deal with this situation. Realizing that this first guess of 4:20 is not the correct answer, since the hour hand does not remain stationary and moves when the minute hand moves, the trick is to simply multiply the 20 (the wrong answer) by $\frac{12}{11}$ to get $21\frac{9}{11}$, which yields the correct answer: $4:21\frac{9}{11}$.

One way to understand the movement of the hands of a clock is by considering the hands traveling independently around the clock at uniform speeds. The minute markings on the clock (from now on referred to as "markers") will serve to denote distance as well as time. An analogy should be drawn here to the "uniform motion" of automobiles (a popular topic for verbal problems in an elementary algebra course). A problem involving a fast automobile overtaking a slower one would be analogous.

Experience has shown that the analogy that might be helpful is to find the distance necessary for a car traveling at 60 mph to overtake a car with a head start of 20 miles and traveling at 5 mph.

Now consider 4:00 as the initial time on the clock. Our problem will be to determine exactly when the minute hand will overtake the hour hand after 4:00. Consider the speed of the hour hand to be r, then the speed of the minute hand must be $12r$. We seek the distance, measured by the number of markers traveled, that the minute hand must travel to overtake the hour hand.

Let us refer to this distance as d markers. Hence the distance that the hour hand travels is $d - 20$ markers, since it has a 20-marker head start over the minute hand.

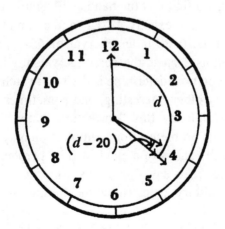

For this to take place, the times required for the minute hand, $\frac{d}{12r}$, and for the hour hand, $\frac{d-20}{r}$, are the same.* Therefore, $\frac{d}{12r} = \frac{d-20}{r}$, and d = $\frac{12}{11} \cdot 20 = 21\frac{9}{11}$. Thus, the minute hand will overtake the hour hand at exactly 4:21$\frac{9}{11}$.

Consider the expression $d = \frac{12}{11} \cdot 20$. The quantity 20 is the number of markers that the minute hand had to travel to get to the desired position, if we assume the hour hand remained stationary. However, quite obviously, the hour hand does not remain stationary. Hence, we must multiply this quantity by $\frac{12}{11}$, since the minute hand must travel $\frac{12}{11}$ as far. Let us refer to this fraction, $\frac{12}{11}$, as the correction factor.

To begin to familiarize yourself with the use of the correction factor, choose some short and simple problems. For example, you may seek to find the exact time when the hands of a clock overlap between 7:00 and 8:00. Here you would first determine how far the minute hand would have to travel from the "12" position to the position of the hour hand, assuming again that the hour hand remains stationary. Then by multiplying the number of markers, 35, by the *correction factor*, $\frac{12}{11}$, you will obtain the exact time, 7:38$\frac{2}{11}$, that the hands will overlap.

To enhance your understanding of this new procedure, consider a person checking a wristwatch against an electric clock and noticing that the hands on the wristwatch overlap every 65 minutes (as measured by the electric clock). Is the wristwatch fast, slow, or accurate?

You may wish to consider the problem in the following way. At 12:00 the hands of a clock overlap exactly. Using the previously described method, we find that the hands will again overlap at exactly 1:05$\frac{5}{11}$, and then again at exactly 2:10$\frac{10}{11}$, and again at exactly 3:16$\frac{4}{11}$, and so on. Each time there is an interval of 65$\frac{5}{11}$ minutes between overlapping positions. Hence, the person's watch is inaccurate by $\frac{5}{11}$ of a minute. Can you now determine if the wristwatch is fast or slow?

There are many other interesting, and sometimes rather difficult, problems made simple by this "correction factor." You may very easily pose your own problems. For example, you may wish to find the exact times when the hands of a clock will be perpendicular (or form a straight angle) between, say, 8:00 and 9:00.

Again, you would try to determine the number of markers that the

*Recall that $r \cdot t = d$, that is, rate times time equals distance, or $t = \frac{d}{r}$.

minute hand would have to travel from the "12" position until it forms the desired angle with the stationary hour hand. Then multiply this number by the correction factor, $\frac{12}{11}$, to obtain the exact actual time. That is, to find the exact time that the hands of a clock are *first* perpendicular between 8:00 and 9:00, determine the desired position of the minute hand when the hour hand remains stationary (here, on the 25-minute marker). Then, multiply 25 by $\frac{12}{11}$ to get $8:27\frac{3}{11}$, the exact time when the hands are *first* perpendicular after 8:00.

For those who want to look at this issue from a non-algebraic viewpoint, you could justify the $\frac{12}{11}$ correction factor for the interval between overlaps in the following way:

> Think of the hands of a clock at noon. During the next 12 hours (i.e., until the hands reach the same position at midnight), the hour hand makes one revolution, the minute hand makes 12 revolutions, and the minute hand coincides with the hour hand 11 times (including midnight, but not at noon, starting just after the hands separate at noon).

Since each hand rotates at a uniform rate, the hands overlap each $\frac{12}{11}$ of an hour, or $65\frac{5}{11}$ minutes.

This can be extended to other situations.

You should derive a great sense of achievement and enjoyment by employing this simple procedure to solve what usually appears to be a very difficult clock problem.

8.7. Where in the World are You?

There are entertainments in mathematics that stretch (gently, of course) the mind in a very pleasant and satisfying way. This unit presents just such a situation. We will begin with a popular puzzle question that has some very interesting extensions, which are seldom considered (but we will consider them later on). It requires some "outside the box" thinking that leaves you with some favorable lasting effects.

Let's consider the question:

Where on earth can you be so that you can walk *one mile south*, then *one mile east*, and then *one mile north* and end up at the starting point?

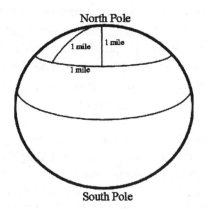

(Not drawn to scale, obviously!)

Mostly through trial and error a clever student will stumble on the right answer: the North Pole. To test this answer, try starting from the North Pole and traveling south one mile and then east one mile. This takes you along a latitudinal line, which remains equidistant from the North Pole, one mile from it. Then travel one mile north to get you back to where you began, the North Pole.

Most people familiar with this problem feel a sense of completion. Yet we can ask: are there other such starting points, where we can take the same three "walks" and end up at the starting point? The answer, surprisingly enough for most people, is *yes*.

One set of starting points is found by locating the latitudinal circle, which has a circumference of one mile and is nearest the South Pole. From this circle, walk one mile north (along a great circle, naturally), and form another latitudinal circle. Any point along this second latitudinal circle will qualify. Let's try it.

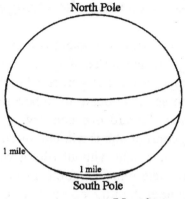

North Pole

1 mile

1 mile

South Pole

(Not drawn to scale, obviously!)

Begin on this second latitudinal circle (the one farther north). Walk one mile south (takes you to the first latitudinal circle), then one mile east (takes you exactly once around the circle), and then one mile north (takes you back to the starting point).

Suppose the first latitudinal circle, the one we would walk along, would have a circumference of $\frac{1}{2}$ mile. We could still satisfy the given instructions, yet this time walking around the circle 2 *times*, and get back to our original starting point. If the first latitudinal circle had a circumference of $\frac{1}{4}$ mile, then we would merely have to walk around this circle 4 times to get back to the starting point of this circle and then go north one mile to the original starting point.

At this point we can take a giant leap to a generalization that will lead us to many more points that satisfy the original stipulations, actually an infinite number of points! This set of points can be located by beginning with the latitudinal circle, located nearest the South Pole, which has a $\frac{1}{n}$th-mile circumference, so that a 1-mile walk east (which comprises n circumnavigations) will take you back to the point at which you began your walk on this latitudinal circle. The rest is the same as before, that is, walking 1 mile south and then later 1 mile north. Is this possible with latitude circle routes near the North Pole? Yes, of course!

8.8. Crossing the bridges

The famous Königsberg Bridges problem is a lovely application of a topological problem with networks. It is very nice to observe how mathematics, used properly, can put a practical problem to rest. Before we embark on the problem, we ought to to become familiar with the basic concept involved. Toward that end, try to trace with a pencil each of the following configurations without missing any part, without going over any part twice, and without lifting your pencil from the paper. For each figure, determine the number of arcs or line segments that have an endpoint at each of the points *A*, *B*, *C*, *D*, *E*.

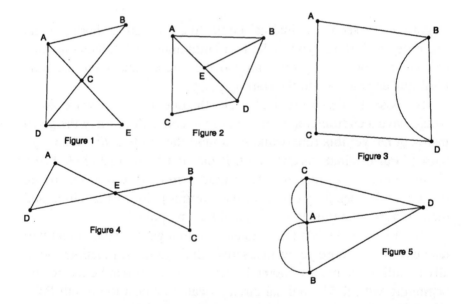

Configurations, such as the five figures above, made up of line segments and/or continuous arcs, are called *networks*. The number of arcs or line segments that have an endpoint at a particular vertex is called the *degree* of the vertex. So if there are three segments or arcs going to a point, that point (or vertex) has degree 3.

After trying to trace these networks without taking your pencil off the paper and without going over any line more than once, you should notice two direct outcomes. The networks can be traced (or traversed) if they have (1) all even-degree vertices or (2) exactly two odd-degree vertices. The following two statements establish this.*

1. *There is an even number of odd-degree vertices in a connected network.*
2. *A connected network can be traversed only if it has at most two odd-degree vertices.*

You might try to draw both traversible and non-traversible networks (using these two theorems).

Network figure 1 has five vertices. Vertices *B*, *C*, and *E* are of even degree, insofar as there is an even number of lines going to those points, and vertices *A* and *D* are of odd degree, since there are three lines going to each of these points. Since figure 1 has exactly two odd-degree vertices as well as three even-degree vertices, it is traversible. If we start at *A*, then go down to *D*, across to *E*, back up to *A*, across to *B*, and down to *D*, we have chosen a desired route.

Network figure 2 has five vertices. Vertex *C* is the only even-degree vertex. Vertices *A*, *B*, *E*, and *D* are all of odd degree. Consequently, since the network has more than two odd vertices, it is not traversible.

Network figure 3 is traversible because it has two even-degree vertices and exactly two odd-degree vertices.

Network figure 4 has five even-degree vertices and can be traversed.

Network figure 5 has four odd-degree vertices and *cannot* be traversed.

To generate further interest, consider the famous Königsberg Bridges problem. In the eighteenth century, the small Prussian city of Königsberg, located where the Pregel River formed two branches, was faced

---

*The proof of these two theorems can be found in A. S. Posamentier and J. Stepelman, *Teaching Secondary School Mathematics: Techniques and Enrichment Units* (Columbus, Ohio: Merrill/Prentice Hall, 6th ed., 2002).

with a recreational dilemma: could a person walk over each of the seven bridges exactly once in a continuous walk through the city?

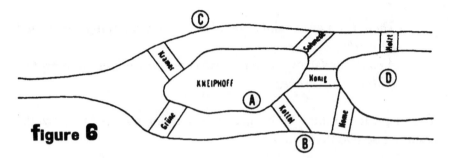

figure 6

In 1735 the famous mathematician Leonhard Euler (1707–1783) proved that this walk could not be performed. The ensuing discussion will tie in our earlier work with networks to the solution of the Königsberg Bridges problem.

We will indicate the island by *A*, the left bank of the river by *B*, the right one by *C*, and the area between the two arms of the upper course by *D*. If we start at Holzt and walk to Sohmede and then through Honig, through Hohe, through Kottel, and through Grüne, we *will* never cross Kramer. On the other hand, if we start at Kramer and walk to Honig, through Hohe, through Kottel, and through Sohmede, and through Holzt, we will never travel through Grüne.

The Konigsberg Bridges problem is the same problem as the one posed in figure 5 on page 266. Let's take a look at figures 5 and 6 and note the similarity. There are seven bridges in figure 6, and there are seven lines and arcs in figure 5. In figure 5 each vertex is of odd degree. In figure 6 if we start at *D* we have three choices: we could go to Hohe, Honig, or Holzt. If, in figure 5, we start at *D*, we have three line paths to choose from. In both figures, if we are at *C*, we have either three bridges we could go on or three lines. A similar situation exists for locations *A* and *B* in figure 6 and vertices *A* and *B* in figure 5. This network *cannot* be traversed.

By reducing the bridges and islands to a network problem, we can easily solve it. This is a clever tactic to solve problems in mathematics. Change the problem situation to a more manageable form.

8.9. the most misunderstood average

Most unsuspecting readers, when asked to calculate the average speed for a round trip with a "going" average speed of 30 miles per hour and a "returning" average speed of 60 miles per hour, would think that their average speed for the entire trip is 45 miles per hour (calculated as $\frac{30+60}{2} = 45$). The first task is to convince readers that this is the wrong answer. For starters, do you believe it is fair to consider the two speeds with equal "weight"? You may realize that the two speeds were achieved for different lengths of time and therefore cannot get the same weight. This should lead you to the idea that the trip at the slower speed, 30 mph, took twice as long and therefore, ought to get twice the weight in the calculation of the average round-trip speed. This would then bring the calculation to the following: $\frac{30+30+60}{3} = 40$, which happens to be the correct average speed.

For those not convinced by this argument, try something a bit closer to "home." A question can be posed about the grade a student deserves who scored 100% on nine of ten tests in a semester and on one test scored only 50%. Would it be fair to assume that this student's performance for the term was 75% (i.e., $\frac{100+50}{2}$)? The reaction to this suggestion will tend toward applying appropriate weight to the two scores in consideration. The 100% was achieved nine times as often as the 50% and therefore ought to get the appropriate weight. Thus, a proper calculation of the student's average ought to be $\frac{9(100)+50}{10} = 95$. This clearly appears more just!

An astute reader may now ask, "What happens if the rates to be averaged are not multiples of one another?" For the speed problem above, one could find the time "going" and the time "returning" to get the total time, and then, with the total distance, calculate the "total rate," which is, in fact, the average rate.

There is a more efficient way and that is the highlight of this unit. We are going to introduce a concept called the *harmonic mean*, which is the mean of a harmonic sequence.* The name *harmonic* may come

---

*A *harmonic sequence* is one where the sequence of the reciprocals of the terms forms an arithmetic sequence (i.e., one that has a common difference between consecutive terms). The reciprocal of the mean of the arithmetic sequence is the mean of the related harmonic sequence and is called the *harmonic mean*. Another way of defining the *harmonic mean* is as the reciprocal of the average (i.e., arithmetic mean) of the reciprocals of the given numbers.

from the fact that one such harmonic sequence is $\frac{1}{2}, \frac{1}{3}, \frac{1}{4}, \frac{1}{5}, \frac{1}{6}, \frac{1}{7}, \frac{1}{8}$, and if one takes guitar strings of these relative lengths and strums them together, a harmonious sound results.

This frequently misunderstood mean (i.e., the harmonic mean) may cause confusion. To avoid this we should recognize that it is used when we are asked to find the average of *rates*. We have a useful formula for calculating the harmonic mean for rates over the same base. In the above situation, the rates of speed we were asked to average were both for the same distance (each a leg of a round trip), which is the same base.

The harmonic mean for two rates, a and b, is

$$\frac{2ab}{a+b},$$

and for three rates, a, b, and c, the harmonic mean is $\frac{3abc}{ab+bc+ac}$.

You can see the pattern evolving, so that for four rates a, b, c, d the harmonic mean is

$$\frac{4abcd}{abc+abd+acd+bcd}.$$

Applying this to the above speed problem gives us

$$\frac{2\bullet30\bullet60}{30+60} = \frac{3,600}{90} = 40.$$

Now let's consider the following problem:

On Monday, a plane makes a round-trip flight from New York City to Washington (with no wind) with an average speed of 300 miles per hour. The next day, Tuesday, there is a wind of constant speed (50 miles per hour) and direction (blowing from New York City to Washington). With the same engine speed setting as on Monday, this same plane makes the same round-trip flight on Tuesday. Will the Tuesday trip require more time, less time, or the same time as the Monday trip?

This problem should be slowly and carefully posed, so that you notice that the only thing that has changed is the "help and hindrance of the wind." All other controllable factors are the same: distances, speed regulation, airplane's conditions, and so on. An expected response is that the two round-trip flights ought to be the same, especially since the same wind is helping and hindering two equal legs of a round-trip flight.

The realization that the two legs of the "wind-trip" require different amounts of time should lead to the notion that the two speeds of this trip cannot be weighted equally since they were done for different lengths of time. Therefore, the time for each leg should be calculated and then appropriately apportioned to the related speeds.

We can use the harmonic-mean formula to find the average speed for the "windy trip."

The harmonic mean is $\frac{(2)(300+50)(300-50)}{(300-50)+(300+50)} = \frac{(2)(350)(250)}{250+350} = 291.667$, which is slower than the no-wind trip. What a surprise!!

This topic is not only useful, but it also serves to sensitize the reader to the notion of weighted averages, a very important concept to remember.

8.10. the pascal triangle

Perhaps one of the most famous triangular arrangements of numbers is the Pascal triangle (named after Blaise Pascal, 1623–1662). Although used primarily in conjunction with probability, it has many interesting properties beyond that field. To better familiarize yourself with the Pascal triangle, it is suggested that you construct one.

Begin with a 1, then beneath it 1, 1, and then begin and end each succeeding row with a 1 and get the other numbers in the row by adding the two numbers above and to the right and left. Beginning this pattern, we would then have the following:

```
            1
          1   1
       1    1+1    1
     1   1+2   1+2   1
   1   1+3   3+3   1+3   1
```

Which is then:

A larger version of the Pascal triangle is shown below:

```
                        1
                      1   1
                    1   2   1
                  1   3   3   1
                1   4   6   4   1
              1   5  10  10   5   1
            1   6  15  20  15   6   1
          1   7  21  35  35  21   7   1
        1   8  28  56  70  56  28   8   1
      1   9  36  84 126 126  84  36   9   1
   1  10  45 120 210 252 210 120  45  10   1
```

In probability, the Pascal triangle emerges from the following example. We will toss coins and calculate the frequency of each event.

Number of Coins	Number of Heads	Number of Arrangements
1 Coin	1 Head	1
	0 Heads	1
2 Coins	2 Heads	1
	1 Head	2
	0 Heads	1
3 Coins	3 Heads	1
	2 Heads	3
	1 Head	3
	0 Heads	1
4 Coins	4 Heads	1
	3 Heads	4
	2 Heads	6
	1 Head	4
	0 Heads	1

Notice that for each number of coins tossed, the number of arrangements represents a row of the Pascal triangle. This lead directly to determining the probabilities when tossing coins. For example, the probability of tossing exactly two heads when four coins are tossed is $\frac{6}{1+4+6+4+1} = \frac{6}{16} = \frac{3}{8}$. Readers should be encouraged to do some investigating of this result by flipping coins and tabulating their results.

What makes the Pascal triangle so truly outstanding is the many fields of mathematics it touches (or involves). In particular, there are many number relationships present in the Pascal triangle. For the sheer enjoyment of it, we shall consider some here. You might try to see if you can locate some others after we consider a few such properties.

The sums of the numbers in the rows of the Pascal triangle are the powers of 2.

$$
\begin{array}{ccccccccccc}
 & & & & & 1 & & & & & & 2^0 \\
 & & & & 1 & & 1 & & & & & 2^1 \\
 & & & 1 & & 2 & & 1 & & & & 2^2 \\
 & & 1 & & 3 & & 3 & & 1 & & & 2^3 \\
 & 1 & & 4 & & 6 & & 4 & & 1 & & 2^4 \\
1 & & 5 & & 10 & & 10 & & 5 & & 1 & 2^5 \\
\end{array}
$$

$$
\begin{array}{c}
1 \quad 6 \quad 15 \quad 20 \quad 15 \quad 6 \quad 1 \quad\quad 2^6 \\
1 \quad 7 \quad 21 \quad 35 \quad 35 \quad 21 \quad 7 \quad 1 \quad\quad 2^7 \\
1 \quad 8 \quad 28 \quad 56 \quad 70 \quad 56 \quad 28 \quad 8 \quad 1 \quad\quad 2^8 \\
1 \quad 9 \quad 36 \quad 84 \quad 126 \quad 126 \quad 84 \quad 36 \quad 9 \quad 1 \quad 2^9 \\
1 \quad 10 \quad 45 \quad 120 \quad 210 \quad 252 \quad 210 \quad 120 \quad 45 \quad 10 \quad 1 \quad 2^{10}
\end{array}
$$

If we consider each row as a number, with the members of the row the digits, such as 1; 11; 121; 1,331; 14,641; etc. (until we have to regroup from the sixth row on), you will find the powers of 11.

```
                              1 . . . . . . . . . .   11⁰
                            1   1 . . . . . . . . .   11¹
                          1   2   1 . . . . . . . .   11²
                        1   3   3   1 . . . . . . .   11³
                      1   4   6   4   1 . . . . . .   11⁴
                    1   5  10  10   5   1  . . . .    11⁵
                  1   6  15  20  15   6   1 . . . .   11⁶
                1   7  21  35  35  21   7   1 . . .   11⁷
              1   8  28  56  70  56  28   8   1 . .   11⁸
            1   9  36  84 126 126  84  36   9   1 .   11⁹
          1  10  45 120 210 252 210 120  45  10   1   11¹⁰
```

The oblique path marked in the next diagram indicates the natural numbers. Then to the right of it (and parallel to it) you will notice the triangular numbers: 1, 3, 6, 10, 15, 21, 28, 36, 45,...

From the triangle, you should notice how the triangular numbers evolve from the sums of the natural numbers. That is, the sum of the natural numbers to a certain point may be found by simply looking to the number below and to the right of that point (e.g., the sum of the natural numbers from 1 to 7 is below and to the right, 28).

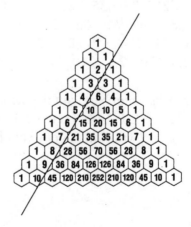

Try to look for the square numbers. They are embedded as the sums (not actually seen) of two consecutive triangular numbers: $1 + 3 = 4$, $3 + 6 = 9$, $6 + 10 = 16$, $10 + 15 = 25$, $15 + 21 = 36$, and so on.

You may also find square numbers (embedded) in groups of four: $1 + 2 + 3 + 3 = 9$, $3 + 3 + 6 + 4 = 16$, $6 + 4 + 10 + 5 = 25$, $10 + 5 + 15 + 6 = 36$, and so on.

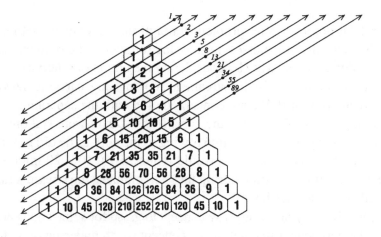

In the above Pascal triangle, add the numbers along the lines indicated. You will be astonished to find that you have, in fact, located the Fibonacci numbers:

1, 1, 2, 3, 5, 8, 13, 21, 34, 55, 89, 144, . . . (see section 1.18).

There are many more numbers embedded in the Pascal triangle. Try to find the pentagonal numbers: 1, 5, 12, 22, 35, 51, 70, 92, 117, 145, . . . (see unit 1.17). The turf is fertile. The challenge to find more gems in this triangular arrangement of numbers is practically boundless!

8.11. It's all relative

The concept of relativity is generally not well understood by most nonscientists. Although it is often associated with Albert Einstein, it has many applications. It may be a difficult concept to grasp for some, so we shall exercise patience and support as we gently navigate further. Begin by presenting the following problem:

While rowing his boat upstream, David drops a cork overboard and continues rowing for 10 more minutes. He then turns around, chasing the cork, and retrieves it when the cork has traveled 1 mile downstream. What is the rate of the stream?

Rather than approach this problem through the traditional methods, common in an algebra course, consider the following: the problem can be made significantly easier by considering the notion of relativity.

It does not matter if the stream is moving and carrying David downstream, or is still. We are concerned only with the separation and coming together of David and the cork. If the stream were stationary, David would require as much time rowing to the cork as he did rowing away from the cork. That is, he would require 10 + 10 = 20 minutes. Since the cork travels 1 mile during these 20 minutes, the stream's rate of speed is 3 miles per hour.

Again, this may not be an easy concept to grasp for some and is best left for them to ponder in quiet. It is a concept worth understanding, for it has many useful applications in everyday life thinking processes. This is, after all, one of the purposes for learning mathematics.

8.12. Generalizations require proof

It can be very tempting to let lots of consistent examples lead you to a generalization. Many times the generalization is correct, but it doesn't have to be. The famous mathematician Carl Friedrich Gauss was known to have used his brilliance at calculating and mentally processing number relationships to form some of his theories. Then he proved his conjectures, and his contributions to the field of mathematics have become legendary. You must be cautious not to draw conclusions just because lots of examples fit a pattern. For example, there is the belief that every odd number greater than one can be expressed as the sum of a power of 2 and a prime number. So when we inspect the first few cases, it works.

$$3 = 2^0 + 2$$
$$5 = 2^1 + 3$$
$$7 = 2^2 + 3$$
$$9 = 2^2 + 5$$
$$11 = 2^3 + 3$$
$$13 = 2^3 + 5$$
$$15 = 2^3 + 7$$
$$17 = 2^2 + 13$$
$$19 = 2^4 + 3$$
$$\vdots$$
$$51 = 2^5 + 19$$
$$\vdots$$
$$125 = 2^6 + 61$$
$$127 = ?$$
$$129 = 2^5 + 97$$
$$131 = 2^7 + 3$$

This scheme worked for each number we tested up to 125, but when we reached 127 there was no solution. Then it continued again. Thus, this cannot be generalized. Caution should be taken before jumping to conclusions, especially when no proof has been developed. This is a good example of drawing premature conclusions. Above all, it is instructive.

Epilogue

f you made it this far without skipping units, then you may have
been won, as a convert, back to mathematics. Even those who
may have skipped some charmers ought to be convinced that
mathematics is beautiful. Demonstrating this was, after all, the
aim of this book. You may recall that I said from the outset that
this book was designed to deflate the ever-popular notion that it is
chic to be weak in mathematics. For no other school subject would
anyone blatantly claim this deficit. We covered the entire spectrum
of elementary mathematics, and for each area we have selected
easily understood examples that would motivate even the most
mathematically damaged in our society. When Gauss referred to

mathematics as the queen of the sciences,* he had not intended that scientists from other disciplines would refer to mathematics as the handmaiden of the sciences, that is, they would judge its value by its usefulness to the other sciences. By this time in your reading of this book, you should have realized that there is much to admire in mathematics in its own right, and not that its primary appeal is its usefulness to other disciplines. Naturally, the latter point is one that keeps mathematics high on the list of important areas of study in our society, but it would be so much more effectively taught and maintained if its appeal could rest only on its own inherent beauty.

The effort to show this beauty in mathematics was done in a variety of ways. First, there are the truly delightful, arithmetically clever processes that have become well-kept secrets and that we have attempted to expose for the purpose of exhibiting other ways of thinking. The quirks in our number system present us with some truly amazing number patterns or almost inexplicable phenomena. They are presented to delight you and to demonstrate that there are some really nice things in mathematics. Further, the completely unexpected connections between various seemingly unrelated branches of mathematics always have great appeal. For example, the many fields of mathematics bridged by such topics as the golden ratio, the Fibonacci numbers, and the Pascal triangle reveal the interconnectedness of this rich discipline. While we're on the notion of the unexpected, the problems presented in chapter 4 show how, with some "outside-the-box" thinking, some problems lend themselves to very clever solutions—the kind of solutions that evoke a "gee-whiz" response, and will hopefully inspire the reader to search for other examples on which to try these unusual techniques.

The chapter on geometry is the one where we can appreciate the visual beauty in mathematics. How surprisingly invariants appear. Of course, these can be best seen by using a computer program such as Geometer's Sketchpad, where a dynamic presentation is possible.

Where possible (and appropriate), historical notes have been pro-

---

*Carl Friedrich Gauss (1777–1855), one of the greatest mathematicians of all time, actually said, "Mathematics is the Queen of the Sciences and Arithmetic the Queen of Mathematics. She often condescends to render service to astronomy and the other natural sciences, but under all circumstances the first place is her due."

vided so that you can put many of these wonderful ideas in a historical context. There is always something appealing when the human element is infused into the discussion of mathematics.

Now that you have been turned on to mathematics (by evidence of having reached this epilogue), you ought to begin collecting books on mathematics, reading them and holding onto them for reference. There are many books on recreational mathematics that can carry you forward as you further pursue your growing love for mathematics. As an ongoing exercise, you might challenge yourself to make a list of subtle applications of mathematics in the daily newspapers. These may be found in a journalist's reasoning, a summary of data presented, the slanting of a story by (mis)use of data, the calculation of data (sometimes incorrectly), or the interpretation of data, which sometimes can be explained in a manner completely opposite to what the writer has done. Back in 1987, as I was reading the *New York Times* with my daughter, we noticed a journalist's error regarding the Pythagorean theorem. My daughter urged me to send a correction to the editors, which I did. Here you can see the article and the response. This then made me a much more vigilant reader of the newspaper. So whenever I see an error, I am now quick to respond. As I mentioned in the introduction, this book is the outgrowth of the almost five hundred letters I received in response to my comments in the *New York Times* (Op-Ed) on January 2, 2002. I would hope that this could be a model for others to read the newspapers and comment where appropriate to keep the mathematics correct. Now fortified with this newly developed love for mathematics, this is the least one could expect.

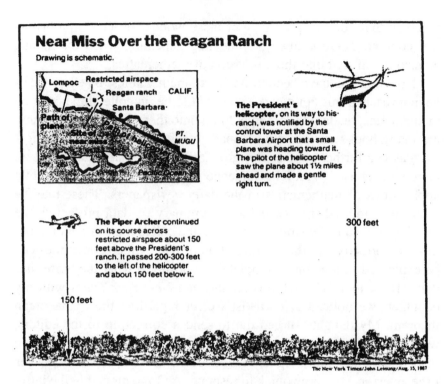

Near Miss Over the Reagan Ranch
Drawing is schematic.

Lompoc Restricted airspace
Reagan ranch   CALIF.
Santa Barbara
Path of plane
Site of
PT. MUGU
Pacific Ocean

The President's helicopter, on its way to his ranch, was notified by the control tower at the Santa Barbara Airport that a small plane was heading toward it. The pilot of the helicopter saw the plane about 1½ miles ahead and made a gentle right turn.

The Piper Archer continued on its course across restricted airspace about 150 feet above the President's ranch. It passed 200-300 feet to the left of the helicopter and about 150 feet below it.

300 feet

150 feet

The New York Times/John Leinung/Aug. 15, 1987

F.A.A. Revokes License of Pilot in Near Miss with Reagan Copter

by Steven V. Roberts
Special to The New York Times

The Federal Aviation Administration, citing a "serious breach of safety," today revoked the license of the pilot of a small plane who flew <u>within 200 feet</u> of President Reagan's helicopter. The pilot was identified as an Army private who was absent without leave. . . . ∎

From the front page of the article, *New York Times*, August 15, 1987.

The Value of Pythagoras in the Cockpit

To the Editor:

Perhaps the most often asked questions of mathematics teachers are, "Why must we learn this stuff?" and "Where are we ever going to use it?" Although the answer is not "to correct the *New York Times*," it is still useful to enable students to read critically, and not just accept information because it is on a printed page.

On your front page Aug. 15, you report that the near miss of President Reagan's helicopter by a private plane over Santa Barbara, Calif., occurred "within 200 feet." Your diagram caption says the plane "passed 200–300 feet to the left of the helicopter," making the minimum horizontal distance 200 feet. With the vertical distance of 150 feet that you show, a right triangle may be formed, whose hypotenuse length is the actual distance the plane was from the helicopter. To apply the Pythagorean theorem ($a^2 + b^2 = c^2$), the one thing most people remember from high school mathematics, this distance is 250 feet—more than "within 200 feet."

This is offered as the sort of thing mathematics teachers (and even parents) ought to point out to students who question the value of mathematics. ∎

Alfred S. Posamentier
Professor of Mathematics
Education and Associate Dean
School of Education,
City College
New York, Aug. 15, 1987

My response from the *New York Times*, August 30, 1987.

Glossary

ALGEBRAIC IDENTITY. An algebraic equation that is true for all values of the variable.

ALGORITHM. A mechanical procedure for solving a problem in a finite number of steps.

ALPHAMETICS. A type of mathematical puzzle in which a set of words is written down in the form of an ordinary arithmetic operation, and it is required that the letters of the alphabet be replaced with digits so that the result is a valid arithmetic statement.

ALTITUDE. The perpendicular distance from a vertex of a polygon to an opposite side.

ANALOGOUS. Having a similarity of some sort, usually applied to sequences of reasoning; for example, "The proof of the second case is analogous to the proof of the first case."

ARITHMETIC MEAN. The value obtained by dividing the sum of a set of quantities by the number of quantities in the set. Also called the average.

ARITHMETIC SEQUENCE. An ordering of terms in which each term (except the first) differs from the previous one by a constant amount.

ARITHMETIC SERIES. The sum of the terms of an arithmetic sequence. One formula is $S = \frac{n}{2}(a + l)$, where S is the sum, n is the number of terms, a is the first term and, l is the last term.

ARRAY. A data structure in which elements are arranged in rows and columns. Frequently referred to as a table.

AVERAGE SPEED. The total distance divided by the total time. Also see harmonic mean.

BASE OF A NUMBER. The number of digits in a particular numbering system.

BICONDITIONAL. Having a codependent relationship; related by an "if and only if" condition.

BINARY NUMBERS. A method of representing numbers in which only the digits 0 and 1 are used. Successive units are increasing powers of 2 (also called "base 2").

BISECTOR. A segment, ray, or line dividing a geometrical object into halves. A line through the vertex and dividing an angle into two congruent angles is called the angle bisector.

CASTING OUT NINES. A method that can be used to check arithmetic by removing the sums of nine.

CENTROID. The center of gravity of a geometrical figure. It is the point at which one could balance the geometric figure on a pinpoint. In a triangle it is also the point where the medians are concurrent.

CIRCUMCIRCLE. A circle containing all the vertices of a polygon. There are some polygons that do not have circumcircles. All triangles have circumcircles.

CIRCUMFERENCE. The length of the entire arc of a circle. Found by the formula $C = 2\pi r$, where r is the radius of the circle.

COLLINEAR POINTS. Points that lie on the same straight line.

COMMON FACTOR. A number that divides two or more given numbers with no remainder.

COMPLEMENTARY. Two angles are complementary if the sum of their angles equals 90°.

COMPLEX CONJUGATE. One of a pair of imaginary numbers differing only by the sign of the imaginary parts. $a + bi$ and $a - bi$ are conjugates, where $i = \sqrt{-1}$.

CONCAVE. Opposite of convex. See convex

COMPOSITE NUMBER. An integer divisible by at least one number other than itself or 1. It is also a non-prime number other than 1.

CONCENTRIC. Sharing a common center.

CONCURRENCY. The property of sharing a common point.

CONGRUENT POLYGONS. Polygons that have exactly the same shape and exactly the same size.

CONIC SECTION. One of a family of curves formed by taking the intersection of a plane and the infinite double-cone. Conic sections include parabolas, ellipses, hyperbolas, and circles.

CONVERGENT SERIES. A series whose sum exists and is finite.

CONVEX. A convex figure is one that contains every segment between points in its interior; opposite of concave figure.

CYCLIC QUADRILATERAL. A quadrilateral all of whose vertices lie on one circle. Both pairs of opposite angles of a cyclic quadrilateral are supplementary (i.e., have a sum of 180°).

DIVISIBLE. Capable of being divided with a remainder of 0.

DUALITY. A correspondence between certain pairs of statements in geometry. Two statements are *duals* of one another when all of the key words in the statement are replaced by their *dual* words. For example, *point* and *line* are dual words, *collinearity* and *concurrency* are duals, *inscribed* and *circumscribed* are duals, *sides* and *vertices* are duals, and so on.

EQUILATERAL POLYGON. A polygon in which all sides have the same length. If the angles are also the same measure (*equiangular*), the polygon is called a *regular polygon*.

EQUIVALENCE. A biconditional, codependent relationship.

FACTOR. In algebra, an expression that divides another expression without a remainder. A number that exactly divides a given number.

FACTORIAL. The result of multiplying all integers from one to the number. For example, "five factorial" (written as 5!) is equal to the product: $5 \cdot 4 \cdot 3 \cdot 2 \cdot 1$. Also 0! is defined to equal 1.

FERMAT'S LAST THEOREM. For all non-zero integers x, y, and z, and for n > 2, there are no solutions to the equation $x^n + y^n = z^n$.

FIBONACCI NUMBERS. The sequence of numbers, 1, 1, 2, 3, 5, 8, 13, . . . in which each successive number, from the third number on, is equal to the sum of the two preceding numbers.

FIGURATE NUMBER. An integer that can be represented by an array of dots forming a regular polygon.

FOUR-COLOR MAP PROBLEM. Any map in a plane can be colored using at most *four* colors in such a way that regions sharing a common boundary (other than a single point) do not share the same *color*.

FRIENDLY (AMICABLE) NUMBERS. Two numbers such that each is equal to the sum of the proper divisors of the other.

GOLDBACH'S CONJECTURE. Every even number greater than 2 can be expressed as the sum of two primes.

GOLDEN RATIO. The proportion of the division of a line segment such that the smaller segment is to the larger segment as the larger segment is to the whole line segment. Golden Ratio = $\frac{\sqrt{5}-1}{2}$ $\approx .618033988$.

HARMONIC MEAN. The inverse (or reciprocal) of the average of the inverses of values. The reciprocal of the mean of the arithmetic sequence (i.e., one that has a common difference between consecutive terms). For example, the harmonic mean of 2, 5, and 10 is $\frac{3}{\frac{1}{2}+\frac{1}{5}+\frac{1}{10}} = 3.75$.

HERON'S (HERO'S) FORMULA. A formula that can be used to find the area of any triangle given the lengths of the three sides. $A = \sqrt{s(s-a)(s-b)(s-c)}$ where A is the area, s is the semi-perimeter (half the perimeter), and a, b, c are the lengths of the sides of the triangle.

HYPOTENUSE. The side of the right triangle that is opposite the right angle. It is also the longest side of a right triangle.

INCENTER. The center of the inscribed circle; the intersection of the angle bisectors.

INTEGERS. Whole numbers, including positive numbers, negative numbers, and zero.

INTERCEPTED ARC OF AN INSCRIBED ANGLE. An arc of a circle formed by the intersections of the two rays of an angle whose vertex is on the circle.

INVARIANT. A value (or relationship) that remains unchanged under some operation.

IRRATIONAL NUMBER. A real number that cannot be represented as the quotient of two integers.

ISOSCELES TRIANGLE. A triangle having two sides of equal length.

LEMMA. A minor theorem that is stated and proved as a part of a proof of a theorem.

LUNE. A crescent-shaped portion of a plane or sphere bounded by two arcs of circles.

MAGIC SQUARE. A square array of numbers such that the sum of the rows, columns, and diagonals are all the same.

MATRIX. A rectangular array, or table, of values.

MEDIAN OF A TRIANGLE. The line segment from a vertex of a triangle to the midpoint of the opposite side.

MULTIPLICAND. In multiplication, the number that is multiplied by another number, the multiplier.

NATURAL NUMBER. An integer, or whole number, that is greater than zero.

OBTUSE ANGLE. An angle whose measure is greater than 90°, but less than 180°.

ORTHOCENTER. The point of intersection of the altitudes of a triangle.

PALINDROME. A number that reads the same backward or forward.

PARADOX. An assertion that is essentially self-contradictory, though based on a valid deduction from acceptable premises.

PERFECT NUMBER. A number that is equal to the sum of its proper factors.

PERFECT SQUARE. A number that can be obtained by multiplying any number by itself.

PERIOD OF DIGITS. A group of digits that repeat.

PIGEONHOLE PRINCIPLE. The property that if $n + 1$ objects (pigeons) are distributed among n categories (pigeonholes), then at least one category will have two objects (pigeons) in it.

POLYGON. Any two-dimensional closed shape composed of three or more line segments, such as a hexagon, an octagon, or a triangle.

POLYHEDRON. A three-dimensional closed shape formed by polygons.

PRIME NUMBERS. Integers whose only divisors are itself and one.

PROPER FACTOR (PROPER DIVISOR). A factor (or divisor) of a number that is not the number itself.

PROPORTION. The equality of two ratios are equal.

PYTHAGOREAN THEOREM. In a right triangle, the sum of the squares of the lengths of the legs is equal to the square of the length of the hypotenuse. Commonly expressed as $a^2 + b^2 + c^2$, for a right triangle whose sides have lengths a, b, c.

PYTHAGOREAN TRIPLE. A grouping of three integers satisfying the Pythagorean theorem. The triple is called *primitive* if there is no number other than 1 that divides each of the numbers in the triple.

QUADRATIC EQUATION. An equation of the form $ax^2 + bx + c = 0$.

QUADRATIC FORMULA. The formula for the solutions of the quadratic equation $ax^2 + bx + c = 0 : x = \frac{-b \pm \sqrt{b^2 - 4ac}}{2a}$.

RATIONAL NUMBER. A real number that can be expressed as the quotient of two integers with denominator not equal to zero.

RECIPROCAL. A multiplicative inverse or the result of dividing 1 by a given number or quantity. For example, the reciprocal of $\frac{a}{b}$ is $\frac{b}{a}$.

RECURSION. The determination of a succession of elements (as numbers or functions) by operation on one or more of the preceding elements according to a rule or formula involving a finite number of steps.

RELATIVELY PRIME. Having no common factor beside the number 1.

SCALENE. A scalene triangle has all sides of different lengths.

SIMILAR POLYGONS. Polygons that have exactly the same shape, but can differ in size.

SUBTEND. To extend under or be opposite to, or to measure off. For example, a chord of a circle subtrends its arc and an angle of a triangle is subtended by the opposite side.

SUPPLEMENTARY. Two angles are supplementary if the sum of their measures is 180°.

SYMMETRY. A state in which parts on opposite sides of a plane, line, or point display the same size, form, or arrangement.

THEOREM. A proposition that can be proved on the basis of explicit assumptions.

TOPOLOGY. The study of geometric forms that remain constant (or invariant) under various transformations, such as stretching and bending.

TRIANGULAR NUMBER. An integer that can be represented as an equilateral triangular array of dots.

TRISECTORS. Rays that divide an angle into three congruent angles.

$\mathcal{I}ndex$

Avon Lake Public Library
32649 Electric Blvd.
Avon Lake, Ohio 44012

Avon Lake Public Library
32649 Electric Blvd.
Avon Lake, Ohio 44012